WELDING

CROWOOD METALWORKING GUIDES

WELDING

RICHARD LOFTING

THE CROWOOD PRESS

First published in 2013 by
The Crowood Press Ltd
Ramsbury, Marlborough
Wiltshire SN8 2HR

www.crowood.com

British Library Cataloguing-in-Publication Data
A catalogue record for this book is available from the British Library.

ISBN 978 1 84797 432 7

Acknowledgements
I would like to thank several people who have helped me tremendously with
this book; without their help it would have been a lot harder than it was. My
niece, Bethany Old, provided the glamour in some of the photographs, which
without her input would have been rather mundane. I think all would agree
she is easier on the eye than a follically challenged, fifty-something guy with
a beard. My son, William Lofting, was a substantial help with the welding
photographs, as he is excellent with a MIG welder. I like to think that I taught
him all he knows about welding, but he makes a better, and more confident,
job of it than I can! He hadn't tried TIG welding until helping on the book,
but now wants an AC-DC TIG welder for his own use. Thanks also go to Pat
Neilson and Michael Davies for the use of their equipment, my wife's aunt,
Audrey Peters, who proof read the text for me, and of course my wife Pam,
who has helped tremendously with keeping morale high when concentration
lapsed. Thank you.

Typeset by Jean Cussons Typesetting, Diss, Norfolk
Printed and bound in Malaysia by Times Offset (M) Sdn Bhd

Contents

Introduction

At one time, not so long ago, if you were looking for welding equipment for the workshop there was little choice: you either plumped for an AC electric arc welder – some more exotic ones giving DC current were available at a price – or it meant an oxyacetylene gas welding plant.

Today things are a little different. Indeed, we are possibly spoilt for choice, for as well as the examples mentioned above we now have MIG welders, TIG welders and Inverter welders, all for reasonable prices with something available to suit most people's budgets.

If I were to say 'Go out and buy a cheap MIG welder and practise on a few scraps of steel', you would possibly be able to stick two pieces of metal together after a fashion, but how would you know if there had been any penetration? To be able to weld properly it will help to have some background knowledge on what is actually happening and, of course, safety concerns come to the fore as in all forms of welding today. Burns are a real risk with all types of welding and all forms of electric welding carry the risk of eye and skin damage due to ultraviolet radiation (*see* Chapter Three).

The first chapter outlines some of the milestones in welding development through the years that have made it possible to produce a good weld with very little effort. Chapter Two is intended to give an idea of what is actually going on in the weld pool itself and the theory behind it, so we can understand what is happening.

All the various welding disciplines will be covered in separate chapters, starting with setting up the equipment to get you going, with step-by-step photographs of the whole process and examples showing the effects of current settings that are too high or too low and other problems. This will enable you to dive straight into whichever chapter is relevant to your welding needs, with all the information that you require to get you going and able to tackle workshop projects as they arise with proficiency and confidence. Specific safety concerns that are critical to each method will be mentioned in the various chapters, but they are covered in depth in Chapter Three.

As with all things practical, while written theory is all well and good, getting someone skilled and proficient to demonstrate how to go about a task and guide you through your first attempts will save you time and frustration as you get to grips with the practicalities of whichever welding discipline you are learning. A friend or colleague may be able to help, but if they have picked up bad habits then these will inevitably be passed on to you and in your ignorance will be perpetuated. The best help and guidance can be found on courses run at local colleges, some of which also provide evening courses to guide you in the correct ways. Health and safety, of course, will be instilled right from the start. When using gas welding equipment, in particular, things can get out of hand extremely quickly, and knowing what to do instinctively in these situations can keep a minor incident from developing into a major one. As stated elsewhere in the book, acetylene can be unstable and in fact is classed as an explosive, but when treated with proper care and attention it is safe to use.

The last chapter will advise on how to choose equipment for your intended purposes and where to purchase it. There is also some discussion of the quantities of shielding gas required and the costs involved.

Throughout the book will be found useful tables on such subjects as welding rod selection and recommended gas pressures. There are also addresses where equipment, consumables and useful advice can be sought.

1 A Brief History

The first evidence for welding dates from not long after the discovery that metals could be extracted from ore by heat. Examples of iron items being hammer welded in a hearth are known from before 1000BC. Under this process the two parts to be joined are heated to just below melting temperature and then quickly hammered together. The extra heat and pressure generated by the hammering enables the surfaces of the two components to fuse and become one.

One fine example of the ancient craft of forge welding is now at the Quwwat al-Islam Mosque in Delhi, to where it was moved at some point in its long history. Known as the Delhi Pillar, it is reputed to have been forge welded, by hand, from several billets of almost pure iron. It stands 23ft 8in (7m) above the ground with a further 3ft (1m) or so buried below, and weighs in the region of 6½ tonnes. It is 16in (400mm) in diameter at its base, tapering upwards to 11½in (300mm) with a fancy finial at the top, although it is

ABOVE: *Forge welding has been the mainstay of welding for centuries.*

LEFT: *During the forge welding process, heat and pressure are used to join the parts as one.*

believed that an ornate figure in the form of Garuda (Sunbird), the Vahana of the Hindu god Vishnu, originally stood on top of the column. According to the Sanskrit inscription at the base, the pillar was constructed during the fourth century AD in honour of the Gupta ruler Chandragupta Vikramaditya. In all the time that the pillar has stood at the mosque, there is little evidence of any rust appearing on the column. While this is a very arid region, it is now believed that this is the result of its phosphorus content, incorporated into the iron from charcoal used in the smelting process. This has caused an extremely thin oxide coating to form, preventing further rusting.

ELECTRIC WELDING

Forge welding and riveting remained the mainstay methods used by blacksmiths to join metal objects until the 1800s. Around this time it was found that carbon electrodes connected to an accumulator (battery) produced an arc. The DC current stored in the accumulators could then be utilized to make a brittle and porous weld. Since the introduction of mains electricity and AC current was still about a century away, recharging would have been carried out with a dynamo, possibly driven by a steam engine or waterwheel.

In 1881 the French engineer Auguste de Méritens was awarded a patent for a method by which the plates of lead accumulators might be welded together with the carbon arc. In 1885 two of his Russian pupils, Nikolai Benardos and Stanislav Olszewski, obtained the first British patent in welding practices. Patent No. 12984 described the method of using the carbon arc with an electric power source to weld metals together. The apparatus used to achieve this was named the 'Electrogefest'. Patents followed in Russia and the USA in 1886 and 1887, respectively.

Development was rapid. Another Russian, Nikolai Slavianov, developed arc welding with an iron electrode in 1888. Similar experiments in the United States led to Charles L. Coffin of Detroit obtaining patents in 1889 for flash-butt welding and in 1890 for spot welding equipment that he had been developing.

The first few years of the twentieth century saw further improvements to these techniques, including the use of hollow carbon rods, filled with metal particles, to act as filler in the carbon arc process. These never gained much popularity, however, although they were later used in specialized processes in the 1940s and '50s. The great leap was made during the 1920s. The effective introduction of thick armour plating during the First World War, for example on warships and battle tanks, led to a demand for

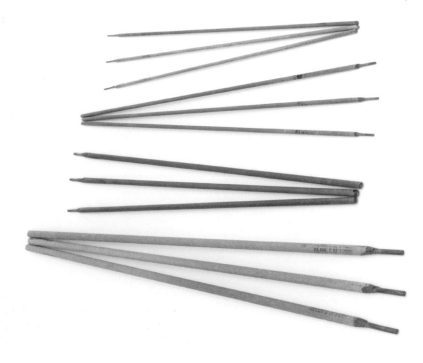

Coated electrodes were designed in the early twentieth century but did not become popular until 1927, when mass production techniques were used to drop the price.

similar armaments. The riveting of such plating, however, was somewhat slow and a faster, more efficient, production method was urgently required. Alternating current (AC) was developed in the early 1920s and power stations were built to supply industry and homes with electricity, although steam power remained the main motive force for many more years.

It was understood that a means had to be found to protect the weld pool from oxygen and nitrogen in the atmosphere is required, since it was this lack of isolation of the weld pool from the atmosphere that made welds brittle and porous. In addition, the welding arc produced by the use of alternating current was shown to be very unstable.

THE COATED ELECTRODE

At the turn of the twentieth century Swedish engineers were developing coatings to cover filler rods: A. P. Strohmenger used clay and lime, while Oscar Kjellberg used carbonates and silicates. The coated electrode was found to do the job. The heat of the arc vaporizes the coating of various clays and silicates into gases that shield the weld pool from the detrimental effects of the atmosphere until the liquid metal beneath has cooled and solidified, leaving a deposit from the remains of the rod coating and any impurities, known as slag, on top of the weld. Coated electrodes were originally produced by dipping lengths of filler wire into a liquid mixture of the coating and setting them aside to dry. The use of these rods did not become widespread until around 1927, when an extrusion process was designed to speed up production, so reducing the price and extending the range of tasks covered.

GAS WELDING

The English scientist Edmund Davy discovered acetylene, a hot burning gas, in 1836. This was followed in 1900 by the development by two Frenchmen, Edmond Fouché and Charles Picard, of an oxyacetylene torch able to create a flame of 3500°C that is ideal for welding. The acetylene obtained for this early development was produced by dripping water on to calcium carbide to release acetylene,

just as early carriage lamps used the acetylene gas liberated by this method to produce a bright white light.

In 1904 Percy Avery and Carl Fisher founded the Concentrated Acetylene Company in Indianapolis to develop ways of storing acetylene as a gas. Acetylene is extremely unstable and, unlike other gases, cannot be directly compressed into an empty cylinder since it quickly becomes unstable at a pressure in excess of around 20 psi and explodes. The cylinder is instead filled with a porous medium, such as balsa wood or asbestos fibre. This is then saturated with acetone, a common solvent. The acetylene gas is very slowly introduced into the acetone and is readily absorbed, like ink on blotting paper, as the pressure is increased. When the pressure in the cylinder is released during burning of the gas, the acetone gives up the acetylene, leaving just the acetone-soaked medium behind in the cylinder, ready for its next charge of gas. Even with the porous medium and acetone in the cylinder, the acetylene gas must be introduced very slowly if it is not to become unstable.

HELIARC PROCESS

From the early days of the Second World War lighter non-ferrous materials, such as aluminium and magnesium, were increasingly used in the new fighter and bomber aircraft and techniques were soon developed to weld these new-fangled materials.

Acetylene was not available in a cylinder until 1913, when Avery and Fisher developed a cylinder with a porous mass inside, soaked in acetone.

Today the engineering world is fairly familiar with the term TIG welding and it is generally supposed that it was developed after the modern MIG welder. Its origins, however, go back to work during the 1920s on using inert gases to shield the weld pool from the atmosphere. It was known as the Heliarc process, after patents taken out in 1941 using helium as the inert shielding gas and a tungsten electrode, as in the modern TIG welder (although nowadays the shielding gas is argon). Modern equipment, of course, is now more compact than the original machines and a lot more versatile.

MIG WELDER

The now ubiquitous MIG welder was developed in 1948. The fixed tungsten electrode in the handset of the Heliarc process was replaced by a continuously fed wire electrode from a roll within the machine. In its initial form it was used for non-ferrous metals, but in the early 1950s it became popular for use with ferrous metals after it was discovered that carbon dioxide – much cheaper and more easily obtainable – made a very suitable shielding gas on steel, although technically it is an active gas and is known as MAG welding (Metal Active Gas). The other main advance was the development of much thinner electrode wires, which made the process more versatile across the range from thin sheet work to heavier sections.

EXOTIC TECHNIQUES

During the 1950s technological advances in the world of welding were being made almost on a daily basis as new ways were developed to produce welds fit for harsher and more extreme environments.

Electron Beam Welding

It was revealed during this decade that electron beam welding was being used in France's growing nuclear power industry, for which good, reliable welds are a necessity, not just desirable. Plasma arc welding, first experimented with in the 1920s, was also developed for industrial use in the 1950s. In this process a stream of gas is heated in a tungsten arc, creating plasma that is half as hot again as a tungsten arc alone can produce. There are many specialist uses and high-grade steels, in particular, benefit from these techniques. The process can also be used for cutting. The plasma cutter itself has now developed into a portable machine that can be used to cut not only steel, but also stainless steel, aluminium, brass and copper. The plasma heats a spot on the surface of the item to be cut and the molten metal is blown clear by compressed air, producing a very tidy kerf, or cut. Oxyacetylene, on the other hand, which has been the principal means of cutting steels, cannot cut metals that do not contain iron, as it is the iron's affinity for oxygen at red heat that allows the cutting to take place. Other applications for the plasma arc, such as metal spraying, have become popular. This technique can be used to give a soft component a harder and tougher outer layer: the outer bearing surface of an engine crankshaft, for example, needs to resist wear, but if the core is too hard the crankshaft might perhaps snap in service.

Laser Welding

The use of lasers for welding has been a fairly recent development. Several individuals and teams were behind the development of the laser. Although Gordon Gould had a working example in 1958, for instance, he failed to patent his device and the first patent was obtained in 1960 by Theodore Maiman. The word 'laser' is an acronym derived from Light Amplification by the Stimulated Emission of Radiation. In very basic terms this describes a substance, solid or gas, being excited by external means. It is this stimulation that coerces the substance into emitting light at one particular frequency, which can then be focused by lenses into a very small beam with extremely high energy levels. This creates a very clean weld with little distortion, which is well suited to use by robots on an industrial production line. This is especially useful in car manufacturing as the laser beam can be directed through fibre optics to the point where the weld is required without requiring the whole machine to be moved as well. The quality of the weld produced is comparable with electron beam welding, but the method is cheaper and has the distinct advantage

The now ubiquitous MIG welder, developed in 1948, became popular for mass car production in the 1950s.

that it is carried out in air with a shielding gas, so protecting the weld pool from the atmosphere. Electron beam welding is much more sophisticated, but needs to be performed in a vacuum.

With experience and improved equipment, what was once a narrow specialist field has now become fairly common knowledge. Welding in the home workshop has never been so easy. The equipment available has become compact, relatively easier to use and, of course, much more affordable. The available choices have never been so good or as varied.

A modern inverter welder, for example, which uses electronic means to alter the voltage/current ratio rather than a heavy transformer, is not much bigger than a lunch box. It comes with DC welding current as standard for mmA stick welding with rutile and low hydrogen rods, and is suitable for welding stainless steel with the correct rods. It is also possible to fit an optional hand torch and regulator for the shielding gas to create a TIG welding set-up, further increasing its scope and versatility for thinner materials. A typical 150 amp model is capable of mmA welding with 4mm diameter electrodes. Together this versatile kit has an overall weight of less than 5kg.

The inverter welder uses electronic control of voltage and current to give DC output. It is light and transportable.

2 Overview of Processes and Ancillary Equipment

A dictionary definition of the word 'weld' usually runs something like 'To unite or fuse' or 'To bring into complete union, harmony etc.'; so to weld something together means to join two parts or more together with a joint that is indistinguishable from its component parts, as it has the appearance and characteristics of the original material. I deliberately have not used the word metal in the last sentence because many other materials can be welded, such as various types of plastics, but from now on I will only be discussing metal welding.

Metals can be joined by many other means, including riveting, screwing, brazing and, of course, soldering. All of these methods, however, are much inferior to true welding with only brazing coming anywhere near close to the strength of a fusion welded joint. Even then it is not a close second: it is in reality just high temperature soldering with the filler material, solder or braze acting as hot glue. Parts to be joined by fusion welding have to be heated sufficiently to

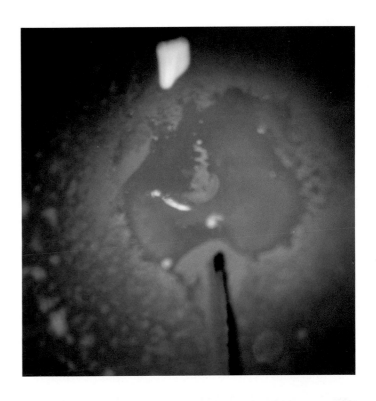

ABOVE: *The weld pool is a molten pool of metal, shielded from the air during welding.*

RIGHT: *A welded joint is a quick and easy way to join metal components.*

Rivets were once the principal means of joining steel plates, used from ships to boilers or, in this case, a portable steam engine.

enable the metal molecules to fuse together either by pressure, as in forge welding, or by adding a filler rod similar in composition to the parent metal to be welded.

In a nutshell, all that is needed to complete a fusion welded joint is a heat source, possibly a filler rod of similar composition to the parent metal and some means – a flux or shielding gas – to exclude the atmosphere from the weld pool while it is still liquid.

Although this book is primarily about welding, a little knowledge of metallurgy and chemistry will help if you are to understand the processes that are going on in the weld pool and the surrounding area. To get a good strong weld we need to exclude those elements in the atmosphere, such as oxygen and nitrogen, that are detrimental to the welding process while the weld is being made. All metals exposed to the atmosphere corrode and form an oxide layer (rust on steel = iron oxide). Whereas the rust on steel carries on oxidizing until all you have left is a pile of rust, aluminium alloys and stainless steels produce an extremely thin but hard oxide layer that prevents further corrosion by excluding oxygen from the surface. At welding temperatures the iron contained in the alloy of steel has an affinity with oxygen in the atmosphere and readily forms iron oxide. As we shall see later, this property can be used to our advantage when cutting steel with oxyacetylene cutting equipment. It is also a disadvantage when welding, though, so the answer is to shield the weld pool from the atmosphere in order to eliminate the

Brazing is a good alternative to welding, but not quite so strong.

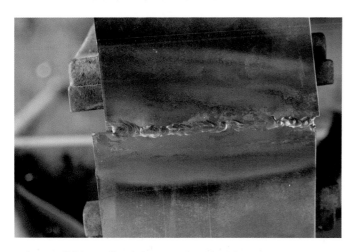

A butt weld is used to join two plates at the edges.

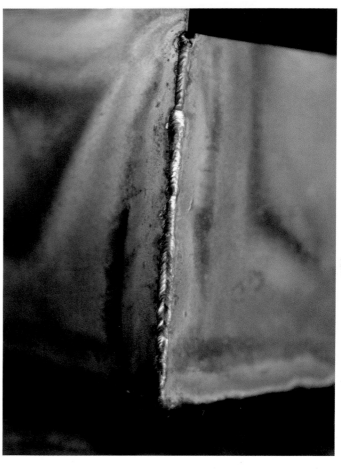

A corner weld, when used with TIG or gas welding, gives a very neat finish without a filler rod being necessary.

A fillet weld joins items perpendicular to one and another.

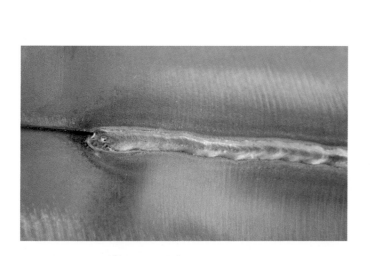

A lap weld is used where two sheets overlap.

The edge weld, used where two edges come together, is good for making fuel tanks.

chance of oxides and other detrimental compounds forming.

It has been established that elements in the atmosphere need to be excluded from the weld pool to produce a good weld. It is just as important that the surfaces of components to be welded are thoroughly cleaned beforehand to remove any rust or other impurities, as it is very possible that otherwise these will be melted into the weld pool, contaminating it with oxides. This will create an inferior weld that is brittle, porous and lacking in strength.

High demand during the early 1970s led to much salvaged steel being recycled and reprocessed, some of which was used in car production plants across Europe. Subsequent reports of engines dropping out of cars, and of jagged holes appearing under the paintwork, were mostly attributed to the quality of the steel produced at the time. Questions were asked about how much reclaimed steel was added to the furnaces and if it was being processed properly. It is now thought that rust on the recycled steel must have been incorporated into the molten steel, rusting through at the most inopportune moment. Although more steel is probably recycled now than previously, a closer eye is kept on its production, ensuring a better quality product.

The most common metal that will require joining in the home workshop is mild steel and, of course, the costs involved must be considered. All initial preparation work in this and subsequent chapters will refer to the preparation of steel items to be welded. Once this has been mastered, in each later chapter we can move on to the additional specific steps necessary

Gap Required for Complete Penetration with Differing Thicknesses of Metal

Thickness of Plate	Gap Width
0–1.5mm	No gap
1.5–2.5mm	Same thickness as sheet
Up to 16mm	Single vee with small root gap
12mm +	Double vee
Horizontal butt weld	15° vee lower, 45° vee upper

for more expensive aluminium and stainless steel preparation.

EDGE PREPARATION

There are five basic types of welded joint, with titles that are fairly self-explanatory: butt weld; fillet weld; lap weld; corner weld; and edge weld.

Butt Weld

The butt weld is produced by welding the two edges together, butted up to one another, with a gap left on thinner material. Single or double bevelling of the edges may be necessary as the thickness increases.

Fillet Weld

The fillet weld can be seen as an inside corner weld, with the items to be welded set at 90 degrees or other angles, depending on the job in hand. For light work the single fillet weld may suffice, but for added strength a double fillet weld can be performed with both sides being welded. More heat input is required than with a butt weld, so an increase of welding current will be required.

Lap Weld

The lap weld is fairly obvious, as the two items to be joined are overlapped. They can be of the same or different thicknesses. The easiest way to weld differing thicknesses is with the thinner on top and most of the heat from the welding heat source concentrated in the lower item, so allowing the edge of the upper piece to flow into the weld pool. The welding rod, or filler wire, should be angled slightly towards the upper item, at approximately 15 degrees from the vertical, if the items being welded are horizontal.

Corner Weld

The outside corner weld is generally easy to produce if the two edges are arranged to meet at one edge, rather than overlapping one edge with the other. If the two items are at 90 degrees this will produce a 'Vee' of 90 degrees, which gives a neat weld after tack welding that needs very little finishing. If the 'Vee' can be set pointing upwards when welding then gravity will assist enormously and, of course, better welds will be produced when the welder is more comfortable.

The hardest part of the corner weld is clamping the components in the correct position before tack welding.

Edge Weld

The edge weld is possibly the simplest technique to produce. It is mainly used in thin sheet work, but on thicker sections can readily be produced with arc welding. Although penetration will be low, it may be useful in low strength situations. The edge weld comes into its own when gas welding or TIG welding, since the weld can usually be produced without the use of a filler rod: as the edges are brought up to fusion temperature, the two amalgamate into one. This leaves a very neat finish, usually requiring just a wire brushing before painting. A gap can form in the weld pool if the two edges are not completely in intimate contact, but this can be filled with judicious application of filler rod before carrying on. A steady, weaving motion of the gas torch or TIG torch will help the two to flow together with very little effort.

GENERAL CONSIDERATIONS

When welding two pieces together, and you require the strength of the original material to be retained, it is not a good idea just to push the pieces together and run a bead of weld across the surface. This would join them together after a fashion, but with

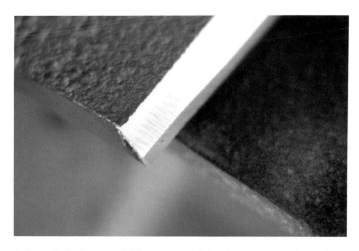

A bevelled edge on thicker material sections ensures complete penetration.

Multiple runs will be required on thicker material to complete the weld: (left to right) first run; second pass; third capping run.

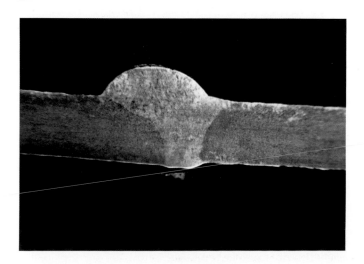

Penetration is the key to a strong weld.

very little, if any, penetration or strength. Penetration is the key word: to gain the strength of the original material, the weld needs to penetrate through the whole thickness of the items being welded. On thin sections an even gap needs to be left between the two edges to be welded. If thicker sections are encountered, then one edge or both need to be bevelled, so that the weld pool can reach down the whole thickness. This will ensure that the fusion process is completed throughout the entire item. On thicker sections it may be necessary to complete the weld in two, three or more passes, leaving the finished weld slightly proud of the surface, in order to strengthen the joint a little. It should be remembered, however, that from a strength point of view the fewer runs that are used to complete the joint, the better, as the localized heating and cooling will have an adverse effect on strength along the line of the weld.

Vertical Welding

Mastering welding techniques while in the horizontal plane is relatively easy, but once off the bench and into some real welding situations, vertical welding will need to be attempted. Vertical down welding is theoretically not as good as vertical up welding, since the impurities congregate in the weld pool as the molten pool travels down, and a high proportion of impurities will be present by the time the bottom of the weld is reached. Vertical up welds will still have gravity trying to pull the weld pool material down, but no more impurities can flow down once frozen, resulting in a more even weld consistency.

Overhead Welding

Whereas horizontal welding has gravity assisting in the filling of the weld pool, overhead welding has the opposite problem, with gravity trying to pull the molten metal from the weld pool. With arc welding, there is a magnetic force helping transfer metal to the weld pool, but with gas welding there is only surface tension helping retain metal in the weld pool. A lower than usual welding current will be needed, as a reduction in the heat input into the weld pool will help retain metal. If there is too much heat, it will be all over the floor (or over you if you are underneath). It can be done, but takes practice.

Heat Sinks

The idea of using a heat sink to remove some of the heat produced while welding would seem counter-productive, as the idea is to heat the metal to welding temperature so that fusion may occur. But when

welding thin or corroded items – thoroughly cleaned, of course – the correct setting for welding the main portion of the weld will be too high for the thin or corroded sections, leading to blow through of the weld and all the remedial actions needed to put this right. The judicial placement of a heat sink below the thin sections will absorb the excess heat and complete the weld without any drama. Another useful job for a heat sink is when making good a component that has wasted away due to corrosion and requires building up, especially around a hole that has to remain to size after the welding. A heat sink to fit the hole will prevent weld metal from filling in the hole, saving time and effort.

It may not appear obvious at first, but if similar material to that being welded is used as a heat sink, there is a very real tendency for the heat sink to be fused together with the component. This leads not only to the frustration of having to right the situation, but possibly the wasting of the component. When welding steel, the usual choice of heat sink material is copper, since it is not only an efficient absorber of heat, but as a non-ferrous metal it will not readily fuse with steel.

Although not necessarily a heat sink, when welding thin aluminium components, which may sag as the welding heat dissipates through the items, a form tool can be made and used to keep the shape and/or positions of the components until they have cooled sufficiently. The best material for this would be stainless steel, as mild steel might contaminate the aluminium during welding.

An assortment of copper offcuts of differing sizes would be a good investment, although these will command high prices as copper scrap prices go through the roof. A good heat sink for use when welding thin car body panels can be made from offcuts of copper tubing used in heating systems; lengths can be purchased from most DIY stores at a reasonable price. In order to use this as a heat sink, flatten the pipe with a hammer so that the two walls of the pipe give a double thickness. This can be clamped under where the welding is to take place and the malleable copper can be shaped to fit the panel being worked on. When welding thicker sections, the relatively thin walled pipe will be insufficient to wick away the heat quickly enough, so thicker sections will be required.

HARDENING AND TEMPERING OF CARBON STEELS

If high carbon steel is heated until it is cherry red, the crystal structure of the alloy will change. If is then quenched, rapidly cooling it, the crystal structure will be frozen and held as it was when red hot. This will have created a steel of extreme hardness, with the downside that it has become very brittle and will snap like a piece of cold chocolate. This structure can be modified by more heat treatment, known as tempering, which modifies the frozen crystals to give more toughness to the steel without losing all of the hardness: the higher the tempering temperature the more hardness is lost, while gaining more resilience. If this is taken too far the steel will become annealed and slow cooling will leave it in its annealed or soft state. It is this process that enables us to create steel tools to cut and work on other steels.

POST-WELDING TREATMENTS

Heat Treatments

As stated above, heating the weld pool area too many times can have an adverse effect on the strength of the weld and the surrounding area. Some more chemistry needs to be mentioned. Steel is an alloy of mainly iron and carbon, along with a host of other materials and metals that can produce a useful selection of differing grades of steels with a wide array of characteristics, ranging from hardness and toughness through to ductility and wear resistance. Mild steel, which generally has no more than 2 per cent carbon, is not altered structurally by heating and cooling, but once above the 2 per cent threshold the steel will be altered in some way.

When welding anything more than mild steel, the heat from the welding can cause cracking in the area adjacent to the weld itself, as the bulk of the steel draws the heat away quickly enough to harden the steel, changing the crystal structure along the side of the weld and creating a brittle line along which cracking is likely to occur. To overcome this, the item can be heated up to red heat and allowed to cool slowly, which not only anneals the whole ensemble but also relieves any stresses built up along the line of the weld.

Peening

Steel, and indeed most metals, can be work hardened by physically working on it by bending

Peening (hammering) will work harden the weld metal to the same characteristics as the metal being welded.

and hammering. Peening describes the hammering of an item's surface with lots of close blows.

This technique was formerly used as a decorative finish to copper items, produced by each hammer blow overlapping the preceding one. The peening process left an attractive dappled finish that also work hardened the metal, giving the items more rigidity and strength.

This technique can be used to good effect along the line of the weld, alleviating stresses in the weld and surrounding metal, and compressing the weld material, which in effect has just been cast from the filler rod, into the weld pool. Peening will close the weld metal structure to a similar composition of the surrounding area. When welding cast iron with nickel welding rods, peening along the weld line will help by compressing the relatively soft nickel as it cools. This should help avoid cracks forming, as the metal contracts on cooling down.

Another scenario where peening is an advantage is when gas welding sheet steel, such as car body panels. The weld deposit left by gas welding is softer than the sheet metal surrounding it. Peening the weld not only helps the metal structure, but strengthens the area as it forces excess weld metal into the panels, rather than making it necessary to grind it off. Welds produced by MIG welding cannot be dressed in this way, as a harder deposit is produced that is similar in composition to the sheet metal itself and will require grinding to bring the weld deposit down to blend in with the surrounding area.

ANCILLARY EQUIPMENT

None of the welding techniques discussed here can be performed properly without other equipment required for preparation and alignment of items to be welded. Once the cleaning, bevelling and other means of preparation have been carried out, equipment such as clamps will be needed to hold things while tack welding and during the welding process itself. Once the welding is completed then more equipment will be needed for cleaning off the likes of slag and welding splatter.

When welding, certain ancillary items of equipment are required to complete the job. Angle grinders, for example, are useful for cutting, grinding and fettling.

Welding Bench

A stout workbench will be a good investment before much welding is undertaken. It does not need to be too fancy, one built from surplus building timber will suffice, and a vice fitted at one end will be a bonus. The workshop may already have a bench fitted, but from a safety point of view it will be better to have a dedicated welding bench. The normal workshop bench is invariably cluttered with tools and part-finished projects, and the usual jumble of oily rags and other flammable substances on or around it is a recipe for disaster. If arc welding on a cluttered bench containing a multitude of metal tools, and something is clamped in the vice with the earth lead clamped to it, the upshot is that placing the electrode holder on the bench, especially if it still has an electrode in it, can cause a cascade of sparks as the current from the electrode finds a path to earth through all the tools on the bench. This may not do much harm, but there is a real danger of fire if the sparks fall on that oily rag. This may be bad enough if it is noticed before it gets going, but if left smouldering after the workshop is shut up for the night, the whole workshop may burn down before the alarm is raised.

Cutting Equipment

The most obvious equipment needed before any welding takes place will be something to cut the materials to be used. The most basic cutting tool, the hacksaw, can be used if only small items are being cut, but monotony will soon set in if used for bigger tasks. Some form of powered cutting tool will then be required.

Power Saws

All manner of powered saws are available. The 'nodding donkey' type power hacksaw, for example, has a reciprocating action and a cut-off switch that automatically switches off the saw when the cut is completed. This has the advantage that the saw, once set, can be left to do the cutting on its own, while you do something else. Along the same lines is the metal cutting band saw, which does all of the above but more efficiently, having a continuous blade in the form of a band. A popular tool is the chop saw, which is similar to the carpenter's chop saw but has a circular tipped blade, or an abrasive disc, that cuts through steel without the cutting lubricant required for the mechanical hack saw and the band saw. The advantage of the power saw, in whatever form, is that angles can be cut when fabrication work is being prepared for welding.

Shearing Tools

Power saws are all well and good for cutting lengths of material, but if sheet work is undertaken then some form of shear will be advantageous. These come in many forms from humble tin snips to the floor- or bench-mounted shear for hand cutting, through to larger foot- or power-operated guillotines for cutting whole sheets.

Tin Snips and Aviation Snips

Tin snips are misleadingly named. They were designed for use by tinsmiths, who extensively employed tin plate, made from mild steel with a protective coating of tin, to make many items that are now made from plastic. Sheet steel, which requires cutting for welding, is predominantly mild steel. Tin snips will cut up to 1.2mm in mild steel and 0.7mm in stainless steel; they will probably cope with somewhat thicker aluminium sheet. Similarly, aviation snips were originally designed for the aircraft manufacturing industry, where it is sometimes necessary to cut complex shapes from sheet material. These come in three forms for straight cutting and left- or right-hand curved cutting. An offset version is available with an offset cutting head that is better for the operator as it keeps the hands clear of the edges, which inevitably will be sharp. Cutting capacity is the same as tin snips.

Floor- or Bench-mounted Shear

Where cuts in larger or thicker sheet material are required, then the floor- or bench-mounted shear will be a good investment. Several types are available. The lightest of these will cut material up to about 1.8mm in thickness. Two discs are fixed in a frame with their sharp cutting edges together. Both discs can rotate, but the top one has a handle attached via a ratchet, so that the top cutting disc rotates on each forward stroke of the handle,

Thinner section material can be cut with tin snips. Be careful of the cut edges.

drawing the sheet in to the blades and cutting it. Curved shapes can be readily produced from this type of cutter. Larger shears have a pair of cutting blades about 150mm long, the lower of which is fixed in the framework attached to the bench or floor. The other blade is pivoted at one end and the other is attached to a cam arrangement that connects it to a long cutting lever, giving the required leverage to produce the cut. As the lever is pulled down the cam multiplies the effort, pushing the top blade past the bottom one in a shearing action. Returning the lever lifts the top blade from the cut and then pushes the sheet manually further into the shear, ready for the next cut. Heavier duty shears are available that will slice through 4mm sheet and have a capacity to cut through 12mm round or square bar.

Oxy-fuel Gas Cutting

If set correctly, oxyacetylene and oxypropane will cut through steel efficiently, creating a cut edge that requires little or no further treatment before welding. Details of the whole process are given later in Chapter Seven. Possibly beyond the exploratory nature of this book is equipment that uses forms of gas cutting to copy a pattern or cut out a circle to produce a disc or hole. If a lot of thick plate is being prepared for welding with bevelled edges, an attachment is available that rolls along the plate edge keeping the cutting nozzle at the selected angle, saving a lot of work.

The bench shear can cut thicker sections of material, as well as some rod and bar sizes.

Colloquially known as the gas axe, oxy-fuel gas cutting equipment makes a quick and tidy job of cutting steel and can be used to cut around a pattern.

Plasma Cutting

Plasma cutting has been around for a while but has recently become a popular method of cutting sheet material, as it will slice through stainless steel, aluminium, brass and copper as well as steel, making the equipment very versatile. It has the ability to follow a pattern, which is useful in a production environment. Advances in inverter technology have produced a plasma cutter that will tackle thick material, while the machine is extremely light and portable.

Angle Grinder

Although the angle grinder and its attachments will be covered in more detail later, it deserves a mention under the cutting tool section. With an abrasive cutting disc fitted, in place of a grinding disc, the angle grinder makes a good cutting tool, if somewhat noisy and dusty. Blades specifically for cutting stainless steel and aluminium are now available, since the standard type clog readily with these materials and quickly lose their efficiency, which can lead to the disc overheating.

While cutting, it is imperative to keep the blade parallel in the cut, as any twisting will stress the reinforcing fibres within the disc, causing the disc to disintegrate. Although the angle grinder will have a guard to protect the operator, as the disc breaks up tangential forces will send lumps of broken disc flying all over the place, hitting the work surface and bouncing back at a dangerous speed. The other danger is that if the disc only partially breaks up, leaving a portion whizzing round still attached to the angle grinder, it will be thrown out of balance and possibly ripped from your grip.

Clamps

Whatever type of welding is to be undertaken, in order to gain a good weld the parts will require some form of clamping arrangement, because as soon as heat is applied the parts will move as the metal expands. With all electrical forms of welding, except for spot welding, the clamps perform an additional task in that they help to maintain a good electrical circuit. Clamps come in all shapes and sizes. The lowly 'g'

The angle grinder fitted with a cutting disc can make straight cuts easier, but goggles are required.

For a successful weld, the items will require some form of clamping to keep them in place while being tack welded.

and 'f' clamps, so called because they look like their respective letters of the alphabet, are tightened down by means of a screw thread. The 'c' clamp has an over-centre locking device, once the size has been set in a way similar to that of 'Mole' grip pliers. For thinner sections, such as car bodywork, there are 'Clekos' in various sizes, spring-loaded edge grips and 'Intergrips'. The disadvantage of the 'Cleko' type of clamp is that you have to drill a hole in the two sheets of metal through which the clamp is inserted, requiring a special insertion tool. You are then left with a hole that has to be welded up after the clamp is removed. The 'Intergrip' type of clamp is used for edge to edge welding. The two parts are clamped together, with the thin blade of the clamp set between the sheets to be joined establishing the right gap between them for correct penetration. Once the two are tack welded between the clamps, they can then be removed and the remaining gaps welded up.

For butt welding thin sheet of car body thickness, 'intergrips' hold the panels in alignment and the correct distance apart for full penetration. These should be removed when the tack weld is complete.

These types of clamps are acceptable if you are welding lap joints or edge to edge, but if the parts to be welded require to be set at an angle, then magnetic clamps come into their own on steel. These come with a preset 90-degree angle at one corner, and 45 degrees at another. Really fancy ones come with an adjustable side, so you can preset your required angle and position them as necessary before welding.

ANGLE GRINDERS

One of the most useful tools in the workshop is the angle grinder. Obviously it can be used for grinding, but with a wire wheel attached it can be put to good

use removing all sorts of surface contamination, from rust through to old paint. It can also be used for sanding, with a backing pad attached. Various grades of sanding discs are available, from the really coarse 40 grit, right through to the fine 180 grit. A fairly modern approach to the sanding disc is the flap wheel, which consists of many flaps of sanding sheet stuck to a backing pad, so that each flap overlaps the previous one around the pad. These are very useful for cleaning up the weld area ready for painting and have the advantage that they last

An angle grinder is useless without its ancillary components.

Magnetic clamps are a boon on steel, as they hold things at the required angle. Fixed and adjustable types are available.

appreciably longer before they need renewing. With a cutting disc fitted the angle grinder makes a very good tool for cutting the metal required for welding. Standard cutting discs are approximately 2–3mm thick, but new types originally developed for cutting stainless steel are much thinner (about 1mm thick), which makes for more efficient cutting. Angle grinders come in a variety of sizes, from the small and light 4in models through to the heavy 9in ones.

The flap wheel is a new take on the sanding disc. It has many overlapping layers, lasts longer than conventional sanding discs and is good for preparing the material for welding. The zirconium grit variety is particularly effective for preparing aluminium.

Comparison of cutting discs of the standard thickness and the thinner version, which gives less heat and enables quicker cutting.

It is the usual trade off: small and light for versatility, or big and heavy for more power for bigger cutting jobs. Some 9in grinders, indeed, are rated at 2,600 watts (more than 3hp).

Wire wheels for the angle grinder also come in several different varieties. Both the straight wire wheel and the cup wire brush may be had in different grades of wire, depending on the severity of the cleaning task at hand. Eye protection is mandatory for all of the above tasks (*see* Chapter Three). This may seem obvious for the grinding, but is just as important for the wire wheeling, as wires fly out of the wire wheel at great velocity. With this in mind it is recommended that a full face visor is used, as these wires do not hesitate to stick in your skin. They will even penetrate through overalls and T shirt, and can be extremely painful.

Air Tools

While we are looking at power tools, it must be remembered that most electrically driven tools have an equivalent available that can be run from an air compressor. The bigger the tool, the more air that is required to drive it. The main advantage with air tools is that the power to drive them comes down the pipe from the compressor in the form of compressed air. In order to gain more power, an electrically driven tool needs a bigger motor built in to it and if you are working at anything other than horizontally on a workbench your arms will soon start to ache. On some jobs you will have to hold up this extra weight at some peculiar angles. While looking at air-driven tools, it is worth remembering that even a small compressor is capable of running tools that require a lot of air for a short time. A spot

When attached to an angle grinder, the wire wheel is very efficient at cleaning and removing light rust.

A spot blast gun will have the area along the line of the weld clean in no time.

blast gun, for example, is a very useful tool but requires a lot of air. Removing paint or rust from along the weld line may take only a few seconds, something a small compressor could probably cope with. You may have to wait for the compressor to catch up on longer runs, but you cannot beat it

for a clean finish and, unlike sanding discs, it will penetrate into the rust pits.

A standard air-blasting nozzle has a very useful purpose when welding thin sheet material. It has two advantages: cooling the weld down rapidly and taking away excess heat, and helping to eliminate

A good selection of air tools to go with your compressor will help with your welding efforts.

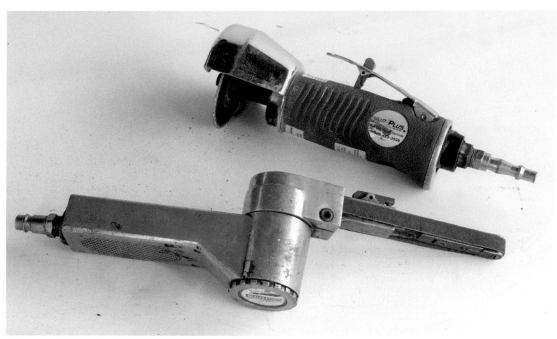

distortion in the panel caused by excess heat. This technique is mainly used with MIG welding. The process involves a short burst of weld, followed rapidly by a blast of cooling air, before repeating the operation. This should prevent the panel from going much above ambient temperature. From the practical stance, it is easier if an assistant is on standby with the air nozzle, as otherwise the heat will have already dissipated into the panel by the time the welding torch has been put down and the air nozzle picked up.

Air tools cover a comprehensive list for the budding welder, with air-driven angle grinders, die grinders, air cut-off saws, percussion tools, such as needle guns, and air chisels, which can be used for cutting and cleaning. The last of these, though, can be extremely noisy. The effect is somewhat diminished with ear protection, but think of your neighbours, who will have little choice but to listen to the racket.

Once all the welding has been completed and cleaned up, ready for a coat of paint, your compressor will be invaluable once again. With a paint spray gun available to run from your compressor, the painting will be done in no time and will look very smart. This, though, is another skill that will need to be learnt.

Anti-splatter sprays are available that help stop weld splatter from sticking, while the cooling compound absorbs excess heat.

Hammers are essential in a welder's tool kit, including (left to right) chipping, ball pein, cross pein, club and panel beating hammers, just a small collection of what is available.

Anti-Splatter Spray

Anti-splatter spray, available in a trigger pack or aerosol can, is very useful for easing the removal of weld spatter after welding and keeping associated equipment clean. The product is water based and is sprayed onto the working area once everything is cleaned and ready to weld. The film left, which is non-inflammable, odourless and non-toxic, helps stop the hot weld spatter sticking to surfaces. A quick squirt into the MIG gas shroud, before welding commences, slows the build-up of splatter, which restricts shielding gas flow, and makes cleaning much easier.

Cooling Compounds

When welding thin or delicate items, or when welding repairs on something such as a car body, where most of the trim remains in place, some means will be required to slow or stop the welding heat from spreading. The traditional compound was made from asbestos fibres, but this has been banned owing to the health risks with asbestos. The modern alternative is made from non-toxic, harmless materials; this putty-like substance is applied as close as is practical to the welding area, keeping the heat contained close to where it is needed and absorbing excess heat as it is drawn into the surrounding metal, before damage is done.

Welding Pliers

Welding pliers are available from many tool outlets specifically for use with the MIG welder. They consist of a pair of long tapered jaws for removing the weld splatter from within the shroud, a cutting edge for trimming back the welding wire, and a notch for gripping the tip when changing the nozzle tips. Whether such a tool is worthwhile is debatable, as a pair of ordinary long nose pliers will do everything that the specialized ones will and they are considerably cheaper. If you already have an extensive tool collection, ordinary wire cutters and a pair of pliers may possibly cover your requirements.

Hammers

A hammer is a very useful tool during welding. As is often the case, once everything is clamped together ready for welding it may be noticed that something has slipped while clamping. Rather than undoing the clamps and starting again, with the possibility that it will slip again on retightening, the easiest option is to give the parts a tap with a hammer, which is usually sufficient to realign things.

Wire brushes come in a variety of designs for cleaning. Stainless steel bristles are essential for use on aluminium.

The traditional method for measuring hammer sizes is in ounces (oz), the usual range from 8oz (250g) through to 32oz (1,000g).

In the engineering world, the usual choice in hammers is to choose a ball pein hammer, with one flat face for ordinary hammering and a ball pein at the other end, which has a multitude of uses from peening through to closing rivets.

Panel Beating Hammers

If you are doing a lot of thin sheet work, such as vehicle bodywork repairs, a set of panel beating hammers will be a good investment. They are often bought in a kit with dollies, for example with the sheet work being dressed with the panel hammer onto the dolly.

Chipping Hammers

A near necessity when arc welding, the chipping hammer is used to chip off the slag coating left on the welding. The chipping hammer has two ends, a pointed end and a vertical knife-edge. The knife-edge is used along the side of the slag run, which should then detach itself, leaving just a little slag to be cleaned off with a wire brush. Any stubborn patches of slag can be worked on with the pointed end, which can get into tight spots and corners more easily.

Wire Brushes

Although rather mundane – and cursed when knelt on – the humble wire brush is an important tool in the welder's kit. It is indispensable for tasks ranging from basic cleaning of rust and crud from the metal surface, to cleaning slag that remains after chipping off with a chipping hammer when arc welding. For quicker and more effective cleaning a rotary wire brush in an electric drill or angle grinder will speed things up, but don't forget eye protection. When working on aluminium it is imperative to use a wire brush with stainless steel bristles, as ordinary steel bristles will contaminate the aluminium.

This is just an illustration of what's on offer to the budding welder in the way of ancillary equipment. There are many other products available, not necessarily at high prices, designed to make welding a pleasure, rather than a chore. Access to these is just a click away, as typing a term such as 'welding tools' into an internet search engine will quickly prove, or you can peruse the tool catalogues produced by mail order tool companies.

3 Safety

Health and Safety has come to the fore in the industrial world today. Every detail and process is put under the spotlight and forensically examined for dangers to human health. This is acceptable as a concept, but it has now reached the stage where the entrance of any large building site has huge signboards displaying, not safety advice, but safety orders that must be obeyed come what may. The consequence of this is that, even in high summer, workers who are not doing anything arduous with their hands wear stout gloves at all times on site for fear of disciplinary measures, up to the sack, due to Health and Safety non-compliance. At the other end of the scale, I have seen a builder using an angle grinder to cut a piece of cast iron guttering to length, without wearing goggles to protect his eyes from the inevitable flying particles.

In your own workshop Health and Safety is down to you. It's your responsibility to decide whether you don the correct safety equipment and where. This all sounds rather 'heavy', but it shouldn't be too onerous a task and everything should soon become second nature. If you are new to the technique or equipment then read all of the information that is supplied. Read a book or go online, where you will find a comprehensive array of literature with all the up-to-date information on the safety aspects of what you are about to do. The Health and Safety Executive (HSE) has a range of downloadable safety leaflets, all free of charge. This will give you a good working knowledge of what is going to happen before you start, and all the possible safety implications along the way.

WELDING DANGERS

The most obvious danger with any form of welding is the potential for burns, not only while welding but for a while afterwards, before everything cools down. With electric welding there is a very real danger from the intense ultraviolet radiation, not just for the person

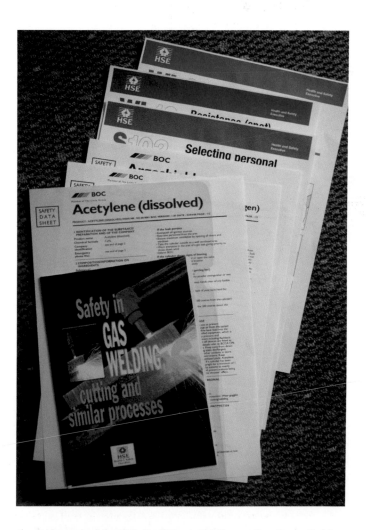

A whole range of safety leaflets and information is available from suppliers of welding equipment and from the Health and Safety Executive.

Perhaps the first of these topics to discuss is where welding should be carried out, from the viewpoint of protecting the workshop from fire, innocent bystanders from burns and radiation, and of course how to protect ourselves.

WHERE TO WELD

For most of us, our workshops cover a multitude of disciplines. Indeed, the title of 'workshop' is perhaps rather grandiose, as in reality we may have to work in the garden shed or garage, doubling up for our workshop, surrounded by house and garden

Safety is paramount and it is essential to wear the correct clothing.

Make sure the welding area is clear of inflammable materials before you start.

welding but for any bystanders and pets. With electric welding and its equipment a further concern is the risk of electric shock.

Secondary dangers associated with welding are fires started from stray sparks from the welding process, which can be started while welding but do not become evident for quite a while afterwards. Paper or rags may smoulder away quietly until enough heat has built up, before bursting into flame, possibly long after the workshop has been shut up for the night. Hot welded parts carelessly laid on a bench, although not red hot, may still be hot enough to ignite paper, cardboard and even the wooden bench itself.

Always make sure that the area you intend to weld in is suitable. This place clearly is not.

Fumes from welding are hazardous to health; ensure that there is adequate ventilation.

The other major concern about welding inside is that fumes and dust particles detrimental to health are given off when welding or preparing to weld. Dust, heavily laden with metal fragments, and particles of grit from the grinding wheel are very bad for the lungs, as are fumes, especially those produced when arc welding, as the flux coating on the welding rods melts and vaporizes to protect the weld pool from the atmosphere. The shielding gases used in MIG and TIG welding, although not poisonous, are asphyxiates and in enclosed spaces they could displace the oxygen in the atmosphere, causing suffocation. Oxyacetylene welding is also not too good for our breathing system: acetylene in high concentrations acts as a mild anaesthetic, and although oxygen is not poisonous at normal concentrations or an asphyxiant, too much oxygen is possibly as bad for the human body as too little. Cutting with oxyacetylene brings with it the dangers of carbon monoxide poisoning, as when the metal is cut the oxygen-cutting stream chemically combines with the red-hot steel, producing copious quantities of carbon monoxide. From the above it follows that if you intend to weld inside it is not only the flammable materials that are a concern. Adequate ventilation is a must, and as we shall see below, it is imperative to choose and use the correct personal protection equipment (PPE) for your safety.

paraphernalia in their rightful place, such as the lawn mower and the garden chairs put away for the winter. As a welding environment, alarm bells should already be ringing: lawn mower – petrol, garden chairs – cushions. All of this is highly flammable, while the shed will probably have a wooden floor. If we are using the garage, the car may have a tank full of petrol, and are there any leaks under the car? If you leave the car outside, sparks when grinding will arc out of the garage door and embed themselves in the windscreen and paintwork. If you have a shared driveway, your neighbour's vehicles may also be in the firing line. Unless you have a purpose-built workshop with an area specifically for welding, the best place to work is outside, away from inflammable substances. This, though, opens another can of worms: can you be observed by innocent bystanders, including children, both the neighbours' and yours? And how about the family dog or cat, or the rabbit in the cage right where you are about to work?

This is more like it – clean, tidy and, most importantly, clear of inflammable materials.

To recap, the area required for welding needs to be free from flammable materials, such as petrol cans, either full or empty, paper or cardboard. There should be plenty of ventilation, so that you and others aren't breathing welding fumes. You also need to ensure that anyone in view of the arc itself is adequately protected from ultraviolet radiation, if necessary by putting up shielding barriers. Proper welding screens designed to filter out the dangerous UV radiation are available for such purposes. Some even come with castors so they can be trundled around to be best placed to give protection where needed.

PERSONAL PROTECTION EQUIPMENT (PPE)

We now need to consider how to protect ourselves. This is where the acronym PPE comes under the

Make sure the collar and cuffs are fastened, to keep out stray sparks.

Some form of overalls is essential as welding can be a dirty pastime; flame retardant overalls are available specifically for welding.

spotlight. Welding can be very hazardous to health, not only that of the operator but also anyone in the vicinity, and means have to be found to eliminate these risks, or at least reduce them to an acceptable level.

Overalls

There are many types of overalls available, from the cheap and nasty through to the fairly expensive. The main concerns are to protect the body from heat, radiation and dirt, but not necessarily in that order. The best overalls for welding are made from cotton or a high percentage of cotton. Nylon overalls are definitely not suitable as the material tends to melt with hot weld spatter and the molten nylon sticks to the skin, causing some nasty burns. Cotton overalls will burn, but they need a sustained heat input for them to actually catch fire. Overalls specifically intended for welding are now available. These have a flame-retardant coating, similar to that on modern furniture, which is self extinguishing to some degree.

Of course it is no good putting on overalls or a boiler suit to find that they do not fasten right up to the neck, and the cuffs really need to be tightly fitting around the wrists to avoid stray splatter finding its way up your sleeves. Practice sessions on the bench won't be too bad, but once you undertake some serious welding away from the bench you may find yourself working at odd angles from which spatter will try to find its way into your clothing. Weld spatter is one thing, but more painful is an errant dollop of weld metal, which is bigger and, holding more heat, burns for longer. Standing up in a hurry, with welding mask in one hand and the welding torch/electrode holder in the other, to shake out the offending hot item is all but an impossibility – so do up all those buttons or zips.

Welder's Leather Apron

If there is not enough protection in a standard pair of overalls, especially when standing at the bench, aprons made from thick leather are available that afford good protection to the front of the body. To some extent the leather is self extinguishing and will take a degree of punishment from a stream of red hot sparks.

Gloves

Hand protection is obviously another necessity, as they are closest to the heat and radiation. If just welding for short periods the lighter 'rigger' type gloves will possibly suffice, but for prolonged periods a pair of stout leather welding gauntlets will be most useful. These cover the cuffs of your overalls and the seams are stitched with Kevlar thread. Gloves specifically for TIG welding are made from a thinner and softer leather, which allows for more dexterity when working on delicate jobs with torch in one hand and filler rod in the other, while still giving the operator's hands the necessary protection from ultraviolet radiation and heat.

For bench work a leather apron will keep sparks and heat at bay, making working more comfortable.

It is not only when welding that hands need protection. Sharp edges abound when any metal is cut. Cutting through steel with an angle grinder invariably leaves a thin sliver as the disc cuts through. Although rather floppy, this will impart a nasty cut if any part of the body is run across it. Thin sheet is possibly the most dangerous from the point of view of cuts to the person when cutting and handling. Stout gloves when handling sheet material will protect the hands from some potentially severe gashes.

Welder's Hat

Sparks produced when working at the bench shouldn't go high enough to cause much worry, but when working at various angles the head may well be in the

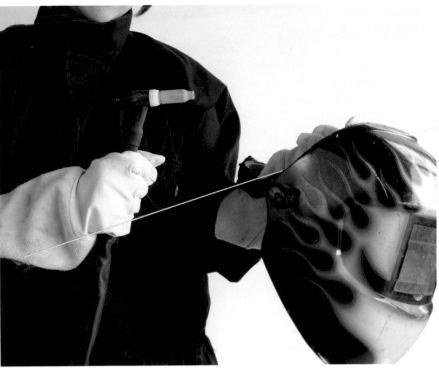

TOP LEFT: *When handling metal with sharp edges, work gloves are an essential piece of equipment.*

TOP RIGHT: *Special gloves made from fine leather are available when more dexterity is required for TIG welding.*

LEFT: *Welding gauntlets should cover the overalls' cuffs. The better ones are stitched with Kevlar thread and protect the hands from the heat and ultraviolet radiation.*

ARC EYE

Arc eye, or ultraviolet keratitis, is the scourge of welders the world over. Exposure of the eyes to ultraviolet light for the briefest of moments will, in some cases, be enough to trigger the effects of arc eye. After glimpsing the initial flash the eyes quickly recover normal sight. It is only later that the full effects come into play, starting with irritation and developing into the painful feeling of having a handful of sand thrown in the eyes, usually as you are trying to sleep. After waiting for three or four hours at your local A&E, if you are lucky, you may get some soothing eye ointment.

Flame-retardant hats and bandanas protect the welder's head and hair from stray sparks. Some are very fashionable indeed!

The basic hand mask.

The full welder's helmet with fixed lens and auto-darkening helmet is much better for serious welding. Auto-darkening helmets are now relatively cheap to buy and are much less hassle to use.

firing line. Those with elaborate hairstyles, or indeed those of us who are follically challenged, may need some appropriate headgear. Specialist welding hats are available, with or without peaks, and bandanas may be bought in flame-retardant materials, but the ubiquitous turned-around baseball cap will do the job, the peak affording some protection to the neck. It goes without saying that you shouldn't use your best one, as it will soon look rather ragged as the welding mask fits over the top and grubby hands will not do it any favours.

EYE PROTECTION

The ultraviolet radiation produced by electric welding is in the UVA and UVB bands. With all that heat being produced, infrared radiation is also in abundance at the other end of the visible spectrum. Exposure of the skin to UV radiation from the sun is bad enough, as past experience of sunburn has probably shown you. The extreme UV from the arc, however, is extremely detrimental to the eyes and will ruin your eyesight in both the short and long term. Arc eye, known to the medical profession as ultraviolet keratitis, is extremely painful but short lived since the cornea of the eye repairs itself fairly rapidly. Longer term damage is caused when the retina at the back of the eye is exposed to the arc long enough to actually burn the sensitive tissues. This process is cumulative: once damaged the eyes never fully recover and repeated burns to the retina will eventually lead to a blind spot in your vision.

With electric welding there is a choice between hand-held face masks, full-face helmets and self-darkening helmets, which have become popular and relatively cheap to buy thanks to developments in liquid crystal technology. As will be seen later, the trickiest phase of welding is the initial striking of the arc. While trying to peer around or under a conventional mask or hand-held or full-face helmet, lining up the welding rod/welding torch before striking the arc is a very fine juggling act. This takes considerable practice to perfect if the eyes are not to get a momentary blast from the arc when the rod accidentally touches the work piece prematurely. This whole scenario is eliminated with the self-darkening mask as the mask stays in place. The viewing lens starts clear, but once struck, the intense light from the arc triggers the lens to darken almost instantaneously to a level safe for the eyes. When the arc is broken, the lens reverts back to its original clear state. This means that if you have a hesitant start there is no need to lift the mask to see where or what you are trying to weld. Another bonus of these self-darkening helmets is that the level of darkness can be adjusted from the standard shade 8 and 9 through to shade 13, which is enough to cover all your electric welding needs, up to 500 amps.

Standard masks have a dark lens comprising two pieces of glass, with a dark membrane sandwiched between them, and are available in different grades. It is a fine balance between shielding the eyes from too much bright light and not having enough light to see by, since when welding the only light available to see what is happening is provided by the arc itself.

I was once welding up a classic MG sports car at a local garage, when a young garage mechanic who was having problems welding a repair for a MOT test on another car, said, 'The MIG welder isn't welding properly'. I had a look and realized the mask lens was too dark: it had a shade 13 lens fitted, which is much too dark for welding car bodywork. I lent him my mask, which had a shade 9 lens fitted, and he started to weld proficiently once again. It turned out that a few days previously he had dropped the mask and cracked the lens. Rummaging in a box in the store, but not knowing that lenses came in various shades, he replaced the broken lens with the first one he picked up, which happened to be a shade 13. After rummaging in the same box I found and fitted a shade 9 lens; he hasn't looked back since.

When gas welding or cutting, the light from the flame, although extremely bright, does not give off any dangerous ultraviolet or infrared radiation, unlike in electric welding. Normal eye protection when using gas equipment is goggles with green lenses fitted to cut the extremely bright glare. A more recent innovation is the full-face shield fitted with a green shade. Although we are principally talking about eye protection from hazardous light emissions produced during the welding process, gas welding has the habit of letting off spectacular pops and bangs when you least expect. These can

Lenses for the welding mask and helmet are available in various shades, depending on the current being used. The replaceable clear cover lens stops the splatter from covering the dark lens.

Arc, MIG and TIG Welding Mask Lens Shades

Lens Shade	Welding Current
9	Up to 75 amps
10	75–150 amps
11	150–200 amps
12	200–400 amps
13	300–500 amps

spray the face with hot particles and can be rather disconcerting for the beginner. Another benefit of the full-face shield is that goggles tend to steam up if you are hot or working in a hot environment, whereas the full-face shield has more access for air to keep the screen clear.

Goggles

It is not only ultraviolet light damage that the eyes have to contend with. Grinding, sanding and wire wheeling all produce grit and rubbish that will cause damage. Goggles come in many shapes and forms

*BOTTOM LEFT: **The view through a No. 13 lens at the same current setting.***

depending on the severity of the task to hand, and in these days of high fashion they are also available in designer styles. Basic goggles with an elasticated band to go around the head will do, although they tend to steam up quickly, usually at the most inappropriate moment. For general work, wraparound goggles are better, with less steaming of the lenses, although offering slightly less eye protection. The full-face visor not only gives eye protection, but offers some protection to the whole of the face, which can be a bonus when used in positions away from the horizontal, protecting the nose and mouth, for example, when lying on your side.

Although the light produced from the gas welding flame is too bright to see what you are doing unaided, it will not incur any eye damage. Gas welding, though, is very prone to spits and splutters.

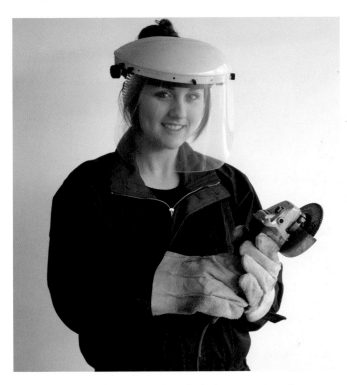

The full face mask protects the whole face, not just the eyes.

A selection of readily available goggles: eye protection is very important.

EAR PROTECTION

Some attention must now be given to hearing protection. It isn't that welding is noisy in itself – it's the grinding and fettling with angle grinders and the like that make most noise. Another reason for wearing earplugs or defenders is to protect the eardrums from burns caused by welding spatter and globules of molten weld metal. This is not too much of a problem when welding on the bench, but once welding from different angles the ears become extremely vulnerable. The odd spatter rattling around inside your ear and singeing hairs can be painful, but make no mistake, a large red hot globule will burn right through your eardrum, which will be detrimental to your hearing for the rest of your life.

Deciding whether to go with earplugs or ear defenders is a matter of personal choice, but when wearing a welding helmet the only option is earplugs. The same goes if you are grinding or sanding with an angle grinder (of course wearing goggles).

BREATHING MASKS

The lungs are another vulnerable area of the body. Just as grit, dust, and metal particles are bad for eyesight, they are also extremely detrimental for the lungs. But these are not the only elements to which the lungs are susceptible. Fumes given off during the welding process itself can be extremely toxic, and the gases used in welding can do harm.

Ears are vulnerable to stray hot sparks as well as to noise. Both ear plugs and ear muffs are available to cope with these.

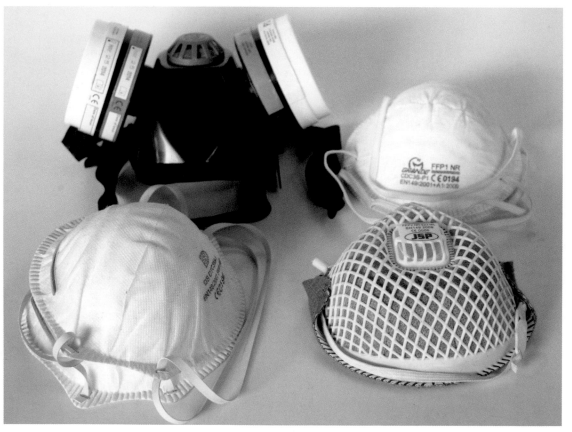

Lung protection from dust and fumes is essential at all times. Here is a selection of masks that are available.

Some Common Sources of Fumes Dangerous to Health

Fume	Source	Symptoms
Aluminium	All alloys containing aluminium	Irritant to respiratory tract
Beryllium	Copper, magnesium and other alloys	Known carcinogen
Cadmium	Stainless steel and plating	Kidney damage emphysema. Suspected carcinogen
Chrome	Stainless steel and plating	Risk of lung cancer, skin irritant
Copper	Brass, bronze alloys etc.	Irritant to eyes, nose and throat. Nausea possible
Fluorides	Electrode coatings	Irritant to eyes, nose and throat. Bone and joint problems
Iron	All steel welding operations	Lung disease Siderosis
Lead	Alloys, solder and paints	Damage to nervous and digestive system, kidneys
Zinc	Alloys, galvanized coatings	Zinc poisoning, sore throat, irritant

This list is by no means comprehensive and many more sources exist; make sure that welding is carried out with adequate ventilation.

In any form of welding, the principle is that the parent metal is being raised to fusion temperature. This inevitably means that the metal is boiling in the weld pool, giving off fumes. As the vaporized metal particles cool in the rising fume column, they will coalesce back into microscopic particles of whatever metal is being welded. The larger particles will sit in the bottom of the lungs, possibly causing or contributing to emphysema. The finer particles will go into the bloodstream. Some elements particularly associated with stainless steel are known carcinogens, notably chrome and some of its compounds. Manganese, which is prevalent in hard facing rods, is known to be associated with Alzheimer's disease.

It is not only the metal fumes that are a hazard. Paint and degreasers left on surfaces being welded can give off or be converted by the heat and radiation into dangerous fumes such as isocyanates and phosgene gas.

The larger particles can be removed fairly readily with a disposable mask. These can be bought in bulk as they are fairly cheap and can be thrown away after the job. It is not worth reusing them as they soon smell stale and after they have been left in the workshop for a bit there is probably as much dust inside them as outside, sticking to the moisture from your breath.

The upmarket versions of the disposable mask are fitted with a one-way valve, which opens on exhaling and lets the air out easily. This has the distinct advantage of releasing your hot breath out of the front of the mask: if you are wearing goggles with the basic type of mask, hot breath escapes from the mask along the top by the side of the nose, which is just the right place for steaming up goggles.

A step up from the disposable type of mask is one with replaceable cartridge type filters. These have a much more robust rubber body that creates a better seal around the face. One or two replaceable cartridges are screwed on the sides. These come in various grades suitable for removing just dust through to vapours, but it has to be said they aren't for use in

This type of mask has replaceable cartridges, which filter out various fumes as well as general dust.

Stout boots with toe protectors are a necessity when working with heavy objects in and around the workshop. However, when welding never tuck overalls into boots.

a confined space, as in this situation the vapours could well have displaced most, if not all, of the oxygen from the air.

FOOTWEAR

Boots, rather than shoes or trainers, should be the order of the day. The hot sparks and spatter showering down all over your feet will soon burn your delicate trainers, and hot particles easily find their way over the top of shoes, making light work of burning through socks and giving you an involuntary dancing lesson. Stout boots come in many forms, although it is essential that they have toe protectors. While much welding related equipment is not necessarily heavy in itself, when dropped from a modest height it will do damage to unprotected

feet. Rigger boots are a favourite for foot and ankle protection, but when welding make sure that the legs of your overalls are not tucked in the top. They should be on the outside to ward off falling sparks and hot globules of metal.

FIRST AID EQUIPMENT

Although precautions have been taken to cover the body from every conceivable angle, it is inevitable that you are going to get burned or cut at some point in your fledgling career as a welder. Everyone I know who has tried welding, cutting or grinding carries a few scars where they have been caught off guard by their equipment or burned by an errant piece of red hot slag. They are mostly experienced engineers and should have known better. I have no medical training, so I am not about to tell you how to cope with a medical situation, but common sense should help and some dos and don'ts may perhaps ease the situation.

Even with care and attention, accidents will still happen in the workshop environment. Make sure there is a first aid kit handy.

carry on working in a dirty environment. If it is more serious, with lots of blood, don't panic: the priority is to stop or slow the bleeding and then seek medical assistance. It is probably a good idea to get someone else to drive you to the A & E department at your local hospital or doctor's surgery, or phone for an ambulance.

It may be prudent to stock your workshop with an adequate First Aid kit, which need not be that expensive. Hang it on the wall and make sure that everyone knows where it is. With all that stuff flying around in the air, an eye wash station would be a good investment, as even when wearing proper eye protection we all get grit in our eyes at times. Rather than poking about trying to remove the offending particle, it's a lot better to wash it out with a sterile solution. If this doesn't work, seek medical assistance. As all these sterile products have a limited shelf life, check occasionally that they're up to date.

FUME EXTRACTION

The dire consequences of welding fumes have already been highlighted. If any serious amount of welding is envisaged in a workshop, then a fume extractor would make a good investment. These draw the fumes from the point of welding via an adjustable collection pipe. The fumes are then vented to air out of harm's way. This is much preferable to using a fan to blow the fumes away, which would simply disperse the fumes to every corner of the workshop.

EQUIPMENT SAFETY

We have comprehensively covered the body protection that is both necessary and available. We will now turn to the basic safety measures that should be observed for the equipment we are about to use.

Logically materials will be prepared in advance of the welding process itself, so safety concerns with the peripheral equipment will be looked at before moving on to the different types of welders and their specific safety concerns.

Burns

The quicker that you can take the heat out of a burn, the less it will hurt. If possible, you should hold the burn either in a bowl of clean cold water or under a running cold tap until the pain eases. If it is bad, cover it with a lint-free bandage and seek medical help. Do not put any creams or potions on it, or listen to old wives' tales about covering it with butter.

Cuts and Grazes

If it's just a nick, make sure it is clean and cover it with a plaster to keep it clean, especially if you intend to

Angle Grinders and Cutters

As we saw in the previous chapter, the angle grinder is, or can be, a dangerous piece of equipment, so personal protection is vital. The official line is that gloves should be worn when using angle grinders, but experience has led me to disagree, I have found that I can hold the tool much more firmly with my bare hands, and the skin on my hands is tough enough to withstand the odd shower of sparks. The occasional wire from a wire wheel sticking into the back of the hand can still hurt, but from the safety aspect, if the grinder slips from your grip it will cause severe damage to you, even death. I vividly remember hearing an ambulance siren and later learning that a local man had been rushed to hospital with a severed jugular vein in his neck. Unfortunately he bled to death before any treatment could be administered. One hand had slipped off when he was using a 9in angle grinder with a cutting disc. The angle grinder was now free to pivot on the other hand and swung up with the inertia from the spinning wheel, cutting through his neck. This sort of thing makes one think, but after a while it is easy to become complacent and even slapdash where tool safety is concerned. The saying 'familiarity breeds contempt' is very apt.

Eye, ear and possibly lung protection are a must, as we have detrimental dust and grit flying about and the angle grinder is extremely noisy in use, causing a nuisance to the whole neighbourhood. It may be prudent to seek a friendly relationship with your neighbours, and possibly let them know that you are going to make a noise. Restrict yourself to short spells, as after only one complaint to the local council you could find yourself being served with a prohibition notice under the latest noise pollution act.

Electric Welders

The mains connection common to all electric welding may be supplied either via a 13 amp plug or by a bigger 16 or 32 amp plug. It is vital that the circuit is protected by an earth leakage trip of some kind, as it takes only about 30 milliamps of current to kill a human being. The mains cable and all the other cables on the welder are vulnerable to cuts and burns. Make sure that the angle grinder has stopped revolving before placing it on the floor. Keep the work area clear, as offcuts with sharp edges will easily cut through the cable's insulation if trodden on. Welding will generate copious amounts of hot sparks, and once completed there will be red hot lumps of metal to contend with.

On the working side, things are much the same. The welding current should not kill you, although the welder may be set to 100 amps or so. If the electrode holder and earth are bridged with your body, you will get a shock possibly strong enough to feel as a tingling sensation, but not enough to kill you. Take my word for it and do not try it, as there is always an exception to every rule. With welding it is all about the current: it's the amps that produce the heat for welding, not the voltage.

The open circuit voltage of an electric welder before an arc is struck is typically in the region of 70 to 80 volts. The reason it does not kill you is that, although potentially there is enough current at this lower voltage, the resistance of the body limits the amount of current that can flow.

The initiation ceremony for new students at the local college was once to stand them all in a circle holding hands. Two students would then grasp an electrode holder or earth clamp to complete the circuit, giving each student a mild tingling from the

OHM'S LAW

With all electrical circuits, the relationship between the voltage and current determines the power available, in watts. For the same power output, the voltage and current are interdependent: lower the voltage and the current will increase, and vice versa. These fundamental relationships were deduced by the German physicist Georg Ohm in the 1820s. The power lines crossing the country that form the national grid operate at such high voltages that the cables can be kept relatively thin. If the grid ran at 240 volts, the cables would have to be astronomically thick to transmit the same power.

current travelling through them. This is probably not allowed today, but it did reinforce to my generation that current was flowing in the circuit, and that it shouldn't kill you.

Since the light emitted from the electric welding process is extremely detrimental to eyesight, make sure your mask has the correct screen in place or your automatically activated mask is set to the requisite setting for the amperage you intend to use. As a guide, the higher the welding amperage, the higher the number shade.

A half-size cylinder of oxygen, used for gas welding and cutting.

Shielding gas cylinders come in various sizes.

Compressed Gas Cylinder Safety

This section covers what is safe practice for cylinders containing shielding gas and oxygen, and what is not. Although they differ in their make-up and respective uses, the potential dangers of gases under high pressure are the same and similar handling rules should be observed. The exception is acetylene, which has a set of rules all of its own.

The cylinders are built to withstand high pressures. The industrial standard pressure of 230 bar is in the region of 3,500 psi. Cylinders are necessarily heavy, so handling requires special care and the use of a trolley, while stout footwear with steel toecaps is essential. Shielding gases – argon and carbon dioxide, or a mixture of both, or in exotic circumstances helium – are all non-flammable but contained under high pressure. When the cylinder is full, 3,500lb (1,300kg) of pressure is attempting to push out every square inch of the cylinder's surface, although this drops as the gas is used up. High-pressure cylinders are formed from a single piece of steel. The only weakness is the thread at the top, where the valve is screwed in and provides the connection for the gas regulator.

Whenever moving a gas cylinder, the valve must be shut and the regulator removed. If the cylinder were to fall over with the regulator attached or knocked while moving, the regulator could shear off the cylinder or the valve block could be ripped out of the cylinder. The principle is similar to that of rockets and other fireworks, although these have a cylinder with a hole at one end and the pressure is produced by chemical reaction. Newton's phrase 'for every force there is an equal and exact opposite' translates, in this case, as that gas escaping uncontrolled from the end of the cylinder will produce force in the other direction, with the cylinder acting like a huge rocket. The same principles apply if the cylinder is dropped, as the valve could be torn from its threads and act as a missile.

Oxygen

The handling of oxygen, from the perspective of it being a compressed gas, is the same as for the shielding gases, although there are dangers with oxygen that need to be addressed. Oxygen itself does not burn, but it supports combustion: in fact, most forms of combustion will only take place in the presence of oxygen or an oxidizing agent (an oxide rich compound). Combustion is a form of chemical reaction that under normal circumstances requires the input of heat to initiate the reaction (burn). Oxygen, however, will react at any temperature with certain materials. From the welding perspective, the biggest danger comes from oil and grease, which will spontaneously combust in the presence of oxygen. *Never use oil or grease on the fittings of any welding equipment*. It follows that oil and grease on overalls could pose a danger, so never hang your overalls on the gas cylinder trolley like a clothes-horse, as a small oxygen leak could spontaneously ignite with disastrous consequences right on top of the gas cylinders. For the same reason oxygen should not be used as a substitute for compressed air.

The acetylene cylinder contains dissolved acetylene gas in an acetone-soaked porous medium. This is also a half-size cylinder.

Acetylene

Acetylene can become unstable if not treated with care (*see* Chapter One). The cylinders differ from others in that they have a porous filling to facilitate storage of the acetylene under pressure. A sudden shock, such as dropping the cylinder or banging it with an object, could cause the acetylene to break down inside the cylinder. Additionally a flashback from the welding torch could cause a shockwave that would travel back up the gas hose to initiate this process. Once this has started, and the cylinder is hot, the only possible action is to call the emergency services.

FIRE PRECAUTIONS

After all this talk of hot sparks from the angle grinder and whatever form of welding we are undertaking, it would be prudent to take fire precautions in the event of an unnoticed stray spark igniting some paper before welding started, or inadvertently placing a smoking hot piece that has just been welded on something flammable. Every workshop should have a fire extinguisher readily to hand, so that at the first sight of flames we can access the extinguisher and know how to use it. At this point it is a bit late to start reading the instructions on how to use your particular appliance. If you can extinguish the fire before it takes hold, all well and good, but if it is already out of control evacuate the area and immediately call the Fire Brigade by dialling 999. Let the professionals deal with the situation. It is one thing to lose your workshop and equipment, but it is entirely something else to be severely burnt or lose your life.

Classes of Fire

Fire extinguishers are available for the three main categories of fires, depending on what materials are alight and what else is around:

Class A: Fires involving paper, cardboard, wood etc.

Class B: Fires involving flammable liquids, such as petrol etc.

Class C: Fires involving flammable gases.

A plain bucket of sand will make a cheap fire quencher before the fire takes hold.

Electrical Fires

Anything that is connected to the electrical supply comes under this heading. The essential rule to remember in the case of an electrical welder or any other electrical apparatus in the workshop is that electricity and water do not mix.

From the above discussion it might appear that, in order to avoid having several fire extinguishers on hand to cover different situations in and around the workshop, it would be sensible to have a powder extinguisher to hand, as these cover a multitude of fires. Let us not, however, overlook the humble bucket of sand or, for that matter, water, which can get the situation under control if used sensibly. A fire blanket could also be a saviour if you inadvertently catch your clothing alight, smothering the flames and starving them of air.

I am resisting trying to sound like Captain von Trapp in The Sound of Music, regimentally ordering his children to perform tasks to the sound of a whistle, but it makes sense for your whole family to know how and when to use your fire-fighting equipment, and where it is located, as extra seconds spent looking for it could make a difference to the situation.

A fire extinguisher is vital equipment in the workshop. Choose one that covers all classes of fire.

The fire blanket is ideal for smothering a fire and can be wrapped around the body quickly if overalls catch fire.

4 Arc Welding

ARC WELDING

Cheap equipment, easy to set up and get you started.

Application
* General fabrication work in mild steel.
* Specialist rods available to join or repair cast iron, sometimes available in small packs.
* Can be forgiving of dirty metal, will tolerate small amounts of surface rust.
* Not suitable for sheet metal work below 3mm thickness.

Also known as mmA (Metal Manual Arc) welding, and colloquially known as stick welding, in this process coated electrodes are used to strike the arc and fill in the weld pool as the electrode melts. The flux coating melts and vaporizes, giving off a gaseous envelope that protects the weld pool from adverse elements in the air.

Arc welding has to be the most simple in terms of equipment and setting up. The basic arc welder consists of a transformer, with a current adjusting facility, one wire with an earth clamp, and another wire with an electrode holder, into which the welding rods are clamped.

If you are completely new to welding, you may perhaps have bought a new machine that comes with the basic equipment to get you started, including rudimentary instructions, possibly a few electrodes, a facemask of sorts, a chipping hammer and wire brush. The first task is to read through the instructions to familiarize yourself with your new equipment.

The arc welder may now seem basic, but it is still very useful, especially for the thicker sections.

Arc Welding Electrode Selection Chart

Electrode Diameter	SWG (Standard Wire Gauge)	Suggested Current
1.6 mm	16	25–50 amps
2.0 mm	14	50–80 amps
2.50 mm	12	80–110 amps
3.25 mm	10	110–150 amps
4.00 mm	8	150–200 amps
5.00 mm	6	200–270 amps
6.00 mm	4	225–350 amps

Current should be selected in the mid range of values given for rod sizes; adjust up or down accordingly as you proceed.

FIRST WELD

Arc welding thin material can be done, but requires a fair amount of practice, so it is probably best to start with something like 3mm plate. Rather than attempting to weld two pieces together from the start, it's a good idea to learn to strike the arc and get a stable weld going, before moving on.

Setting Up the Equipment

Once you are familiar with the instructions, you can get the equipment ready for use. If the leads come as separate items they will need to be connected to the welder. The current should be set for the size of rods to be used and the thickness of material to be welded.

The earth lead needs to make a good electrical connection to the work piece, so any paint or rust should be removed before the clamp is positioned, as a poor earth connection will lead to unsatisfactory welds. It has to be remembered in any form of electrical welding that striking the arc is completing the electrical circuit, so it follows that a poor earth connection will impede the process.

For those just starting out in arc welding, a complete kit can be bought very cheaply, with most of the things you need to get you going.

Striking the Arc

Possibly the hardest part of arc welding to get to grips with is getting the arc to strike. Once this is mastered, the rest is just practising and applying the techniques you have learned to the job in hand. The whole process of electric welding relies on producing and maintaining a complete electrical circuit, the arc being a vital part of that circuit. New electrodes have a chamfer at the end so that part of the rod itself is exposed. Here the rod just needs to be brushed in short strokes against the object to be welded at the start of the weld. If successful, this will start the arc without too much fuss.

Once the rod has been partially used, a tapping action needs to be adopted, as on finishing the previous weld the flux coating on the rod will have melted and covered the end of the rod with slag. This is hard and brittle, just like glass, and needs chipping off before the arc can once again be established.

To initiate the arc, the welding rod needs to be scratched along the surface.

The ends of the electrodes are chamfered to facilitate easy striking of the arc.

If the chipping action is too severe, the flux coating will sometimes crack off, resulting in an unstable arc. Without the necessary flux, there will then be a porous portion of weld until this is burnt off and the rod is solid once more. If this happens the easiest option is to have a waste piece of plate ready for just such an occasion. Strike the arc on this piece until the arc is stable and then return to your work piece. Many welders use the earth clamp to strike the rod on and burn back to a good flux-covered rod. This is not good practice, though, as the clamp ends up encrusted in spatter and sooty deposits, which are likely to fall under the clamp

Once the rod has been partly used, the tip of the rod burns back into the flux coating, making further use easier.

as it is tightened to the work piece, reducing the efficiency of the earth return path.

Sometimes, if the current is set too low for the size of rod, the rod sticks to the work piece as the arc is struck. If you are lucky, you can wiggle it free and attempt another go. It often happens, however, that the rod sticks fast and after a few seconds it is glowing cherry red. The quickest way out of this situation is to release the rod clamp on the electrode holder, but don't forget it will still be very hot and could burn right through your gloves, giving you a nasty burn. This sticking is a sign that the current is set too low, and a slight increase in the amperage of the welder will remedy the situation.

If you have any difficulty in striking the arc, the rod can be struck first on a piece of spare plate. Heating the rod tip makes it easier to strike.

The angle of the welding rod should be about 60–70°, with the rod pointing towards the finished weld.

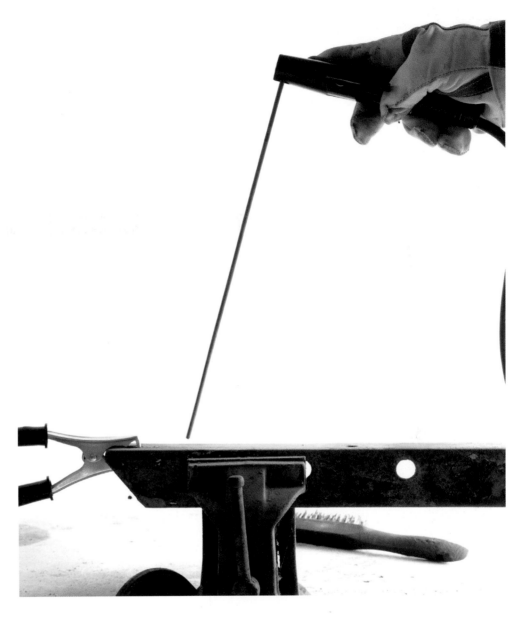

Angles

Now that you have mastered striking the arc and practised running a bead of weld onto a piece of plate, it is time to try a welded joint. Perhaps you should have been thinking about the angle of the rod when practising, but concentrating on positioning the rod before attempting to strike the arc, at the same time as pulling down your face shield, then striking and maintaining the arc, and moving the rod along, is more than enough at one time.

When welding horizontally, the rod needs to be maintained at an angle to the surface being welded of approximately 75 degrees, in line with the joint to be made, and facing in the direction of the weld. In the case of a right-hand weld this would mean moving from left to right; if welding in the other direction, from right to left, then everything will be reversed. This is best practised with the machine switched off, and possibly with your eyes shut, until it is almost second nature. Unless you are using an auto-darkening helmet, once the power is on and the visor is down you will be in the dark until the arc is struck and stabilized.

The arc is extremely bright, so make sure the right eye protection is used.

During arc welding the flux coating on the rod creates a slag covering on the weld. This is removed by chipping or it will peel away by itself if the welds are good, leaving a smooth weld. In this case the bubble at the end was due to rust on the surface.

The Arc Itself

Once struck, the arc length is critical to the welds produced. Too long an arc will not produce the required penetration and lead to an excess of weld splatter all over the place. Too short an arc may possibly result in the weld pool blowing all the way through the article being welded. Somewhere between the two extremes is ideal, but how do you find the ideal arc length? One way is to look at the welds produced: widespread weld splatter indicates that the arc is perhaps too long or the current is too high. Try shortening the arc and turn the current down a little. A good indicator of the welding quality is the slag: if

the weld is good then the slag will start to peel off the weld as it cools and contracts, leaving just the residue to chip off and clean with a wire brush. While cleaning the slag from your welds it is imperative that you wear eye protection. The slag is hard as glass and striking it with a chipping hammer produces lots of sharp splinters that travel in all directions. The same goes for wire brushing, as all the small slag particles will be flicked towards you when brushing along the weld line, so make sure you have your goggles on.

Once you can strike and maintain the arc it is time to do some real welding. You should start by running a bead along a piece of plate before venturing into an actual welded joint.

Tack Welds

Although clamps have been mentioned earlier, tack welds are an essential part of welding. The whole point of a tack weld is to hold the components in position before the main welding run. Components lightly clamped are likely to distort and warp from the full heat of the arc welding. The tack weld should be a short proper weld, rather than a blob of weld that has not properly fused to the components. It should be strong enough and fused, so that if any slight realignment is required with a hammer the welds bend rather than crack off through lack of fusion. If the tack welds crack off, the only remedy is to clean the components, removing all the blobs of weld, slag and splatter, and start again.

The tacks need to be placed regularly along the weld line. Once made, the slag should be thoroughly cleaned from the tack welds, paying particular attention to the ends of the weld, as slag left here will possibly result in the dreaded slag trap, the bane of arc welders worldwide.

Weaving

Any budding welder will be aware that at some point the rod is supposed to be weaved during arc welding. Where, when and how affects the outcome of the weld produced. In arc welding, though, the actual weld pool cannot be seen directly, as it is covered in a layer of molten flux necessary to stop oxidation. Due to the effects of surface tension, the molten metal within the weld pool forms a shape between the two parts being welded that resembles an onion. The weave serves two purposes: it distributes the heat produced at the end of the welding rod evenly between the components being welded, and it adds the filler metal around the edge of the weld pool. In an extension of the onion analogy, it can be seen as adding another layer as the weld moves along the joint. Apart from improving the finished welds, weaving can be used to enhance the weld. Pausing momentarily at each side of the weave will give the filler rod a chance to fuse at the edges of the weld pool and create a slight build-up of filler material, preventing any undercuts that would inevitably weaken the finished weld. When performing vertical welds, where gravity is working against you, correctly weaving the rod will help to overcome gravity and build the weld where required.

Slapdash weaving will give slapdash results in the finished welds. A controlled and paced weaving action leads to superior results, leaving an

Tack welds are important, but they should not be too big. If necessary they can be slightly realigned with a hammer.

even pattern that not only looks good but offers assurance that the weld has been completed with full penetration. Irregular weaving increases the risk of slag traps forming.

Slag Traps

Slag traps not only spoil the look of the welds, but also weaken and make them porous as they go right through the weld. There are two remedies for slag traps. The more effective way is to grind out the offending parts of the weld with an angle grinder and to remake the section or sections of weld. The second is to attempt to pick out all the trapped slag with a pointed object. A small centre punch reground to a shallow angle is quite good for this, used with a hammer to crack and loosen the slag. Once the area is clear of slag, a re-weld can then be attempted. As the area is short and confined, a quick hot weld will be required to avoid a large weld deposit. This can be achieved by going down a rod size and cranking up the current, resulting in a deep weld pool intended so that any remaining trapped slag can be floated up and out.

ELECTRODE CLASSIFICATION

To eliminate any errors when selecting the right electrode for the job in hand, a standard welding rod code has been established. This code tells you what the rod is suitable for, the characteristics of the weld produced, and whether it is suitable for AC or DC current, or indeed both. The code is printed on every rod as well as the boxes they come in: E6013, for example, denotes a rutile coated rod intended for general-purpose welding of mild steel and suited to all welding positions. The box label also carries all the information that you are likely to need, such as the arc striking open voltage (OCV) and drying information for if the rods are susceptible to damp conditions, especially if they are low hydrogen rods, such as E7018.

Code Explained

The code itself is easily explained. The first two digits of a four digit code, or the first three digits of a five digit code, describe the tensile strength of the weld produced in 1000 psi: E60XX describes a rod

A slag trap is the universal bane of arc welders. The only remedy is to chip or grind out and reweld.

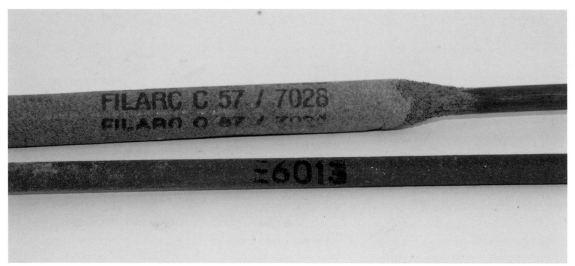

All arc welding rods carry a code to identify the type once out of the box.

capable of producing welds to 60,000 psi tensile strength, and E70XX describes a rod capable of producing welds with 70,000 psi tensile strength. The last but one digit describes the positions that the rod can be used in: EXX1X describes a rod that can be used in all welding positions and EXX2X describes a rod suitable for flat and horizontal welding positions. The last two digits together describe the flux coating on the rod, the type of current the rod is suitable for use with, and the

amount of penetration that the rod will produce: EXX13 describes a rod suitable for all position welding, with a rutile coating suitable for AC or DC current. In conjunction with the tensile strength information, E6013 thus describes a rod suitable for general-purpose welding on mild steels, with medium penetration. The last two digits have a range from 10 through to 28, with the higher numbers indicating those with more flux on the rod. In addition, the flux contains varying degrees of iron

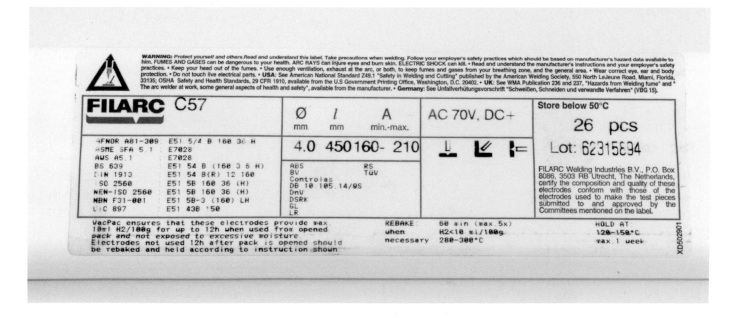

Full details of welding rods and their uses are displayed on the box in which they come.

Common Arc Welding Electrodes Showing Code Differences

Rod Code	Use	Penetration	Comments	AC-DC
E6010	General purpose	Deep	Good on dirty metal	DC
E6011	General purpose	Deep	Good on dirty metal	AC-DC
E6013	General purpose	Medium	Superior bead appearance	AC-DC
E7018	Low hydrogen	Medium	X ray quality weld possible	AC-DC
E7028	Low hydrogen	Medium	As above heavier deposit	AC-DC

powder, which gives these rods a high deposition rate. On more specialist rods, the normal code is followed by a suffix, giving specific information about alloying metals contained in the rod: a suffix of A1 signifies that the rod contains 0.5 per cent molybdenum (Mo); B1 through to B5 indicates the amounts of chrome (Cr) and molybdenum added; C1 to C3 describes the amount of nickel, molybdenum and chrome contained in the rods; D1 and D2 describe the amount of molybdenum and manganese (Mn), and so on.

SPECIALIST RODS

It is possible to weld various grades of steel with the correct welding rods, and in some cases by changing current from AC to DC current, as when welding stainless steel. Rods are available for welding dissimilar metals and this is important for machinery used in mining and earth moving, which can wear at an incredible rate, especially at the points in direct contact with the rock or soil being dug out. Rods have been produced from an alloy of various steels that have extremely high resistance to wear and at the same time are incredibly tough for building up worn parts. Other rods that produce a very hard deposit are used for hard facing applications. Rods made from almost pure nickel will weld broken cast iron without pre-heating the whole casting, provided the metal is clean and only a short length is done at a time, waiting for it to cool before doing another small section.

Open Circuit Voltage

Open circuit voltage (OCV) is the measured voltage between the electrode and the earth of the welding machine before the arc is struck. Welding electrodes have a minimum value at which the arc will strike without fuss; for general-purpose electrodes this is usually 50 volts. The minimum OCV is always listed on the packaging label, but a higher OCV in the region of 80–90 volts is usually specified for some specialist rods. The simplest arc welders produce the minimum voltage for general welding, but if hard facing is to be part of your welding requirements then a set with a higher minimum voltage will be required. Some more sophisticated welding sets come with another voltage tapping on the transformer, giving a higher OCV for more specialized welding.

Repairing Cast Iron

Cast iron is notorious for being hard to repair, but it is always worth attempting to save the casting, if at all possible. The biggest problem with cast iron, for example in engine blocks and gearboxes, is that when cracked it will invariably soak up the fluids that it is supposed to contain. Ideally every trace of oil and grease should be removed. This will be a problem when a large casting needs cleaning as it may not be possible to place it in a solvent to remove the offending material, but something such as brake and clutch cleaner can be washed over the crack to remove it. As with any crack repair, the end of the crack needs to be located. This is important as the root of the crack needs to be removed by drilling through with a small drill. If this is not done you can be sure that, sooner or later, the crack will continue to extend itself once the repair is completed. After the root of the crack has been dealt with, the whole length of the crack will need to be 'veed' out with an angle grinder or carbide burr. This not only gives room for the filler material and helps with penetration, it will remove contamination from the edges of the crack. Ideally the welding should be done in very short bursts of no more than about 20mm, so as not to overheat the casting. Allow the casting to cool down to body temperature before doing the next section, and continue in this way until complete. While waiting for the casting to cool it is advantageous to peen the weld metal as it contracts, which helps prevent cracks from forming between the casting and weld metal interface.

Hard Facing

The heading says it all. When hard facing items belonging to earth moving and impacting equipment, it is the surface that we are building up. From necessity, depending on the actual application, the deposit laid down is wear resistant, impact resistant, or a varying combination of both. Welding rods are available for specific tasks from suppliers. It is always a good idea to discuss your needs with a specialist supplier, as they have all the technical information about the rods they stock, as well as feedback from customers on the performance of these products.

Normal welding – whether arc, MIG, TIG or gas – is all about penetration for the weld to be successful. With hard facing the reverse is more like the truth. Too much penetration of the surface being built up will have the effect of diluting the properties of the electrode being used with the parent metal beneath. Good adhesion is required. When hard facing a softish component, it will be advantageous to have a buffering layer of an intermediate grade rod to help prevent cracks forming between the soft and the hard materials. Pre-heating of the component, with slow cooling after the hard facing, will also help reduce the risk of cracking.

Gouging

Gouging rods are available designed especially for gouging. These have a hard outer flux cover to concentrate the stream of gas produced from the flux in the heat of the arc. Once the arc is struck with these rods, the rod is tilted over to a shallow angle to the working surface in order to allow the stream of gas to blow the molten metal clear, using a sawing action with the rod all the while to maintain the arc. These rods can be very useful for cleaning the underside of a root run when welding thick plate with multi passes, before running a sealing bead of weld. Another use for these rods is for cutting holes in plate. After striking the arc, the rod is kept vertical to the surface of the plate until the molten pool drops through. The size of the hole, once started, can be increased by running around the periphery of the hole with the gouging rod until the approximate size is reached. Although a rather crude method, it is quite effective and quick, even if the precision leaves something to be desired.

ROD CARE AND STORAGE

For successful welding with flux-covered electrodes, it is imperative that the rods are kept away from dampness. Some suppliers of welding rods supply their rods in plastic cartons. These are ideal as they are hard-wearing in a workshop environment and, above all, are waterproof. It makes sense to get in the habit of replacing the carton lid after withdrawing the number of rods that you need for the job in hand. Electrodes left out in the workshop will readily absorb moisture. This will not only have a detrimental effect on the physical structure of the flux coating, but will produce poor quality welds below specification, especially with low hydrogen rods, where welds need to be produced to a certain specification.

Rod Ovens

Specialist welding rod ovens are available for industrial use, where the more specialist rods need to be kept completely free from moisture, before and during use.

Welding rods come in many different types of boxes; the best are made from plastic to help keep moisture out.

If welding rods are purchased in cardboard cartons, a good place to keep them free from moisture until ready to use is in the airing cupboard.

WELDING GENERATOR SETS

Electrical generators can be used to power most types of electrical welding equipment where there is no mains supply, although the average generator has trouble coping with the sudden power surges associated with striking the arc. Welding generator sets are produced specifically for the purpose of arc welding in the field. As long as you have fuel in the generator's engine fuel tank, then welding can continue with the generator controlling the driving engine's governor to maintain the preset current selected.

A welding generator set is extremely useful out in the field, away from mains electricity.

ARC WELDING PITFALLS

Slag requires removal after welding.

Slag traps weakening weld.

Electrodes comparatively expensive compared with MIG welding filler wire.

Special rod ovens required to keep electrodes free from dampness.

High level of fumes produced.

Although arc welding has been superseded by MIG welding in most areas, it still has a place in heavy construction work, especially outside where wind can play havoc with shielding gases. The basic operation is shown below as an example of the technique.

TOP LEFT: *Although arc welding is more tolerant than other types of welding, cleaning rust and rubbish from the surfaces will be beneficial to the end results.*

TOP RIGHT: *Once the metal thickness is above 3mm, the edges should be bevelled to allow penetration.*

BOTTOM RIGHT: *Clamping the parts is an important stage of the operation before tack welding can begin.*

TOP LEFT: *Once tack welded and the slag has been cleaned, weld the joint together, holding the rod at the desired angle and feeding in the rod to keep a constant-sized arc length.*

TOP RIGHT: *As the welded joint cools, the slag covering should start to peel off. Clean it thoroughly with a wire brush to remove all traces of slag. Goggles must be worn, as slag fragments are as sharp as glass. The bottom of the weld shows a slag trap, possibly caused by insufficient cleaning of the tack weld.*

BOTTOM RIGHT: *The underside, showing that the penetration had been good.*

5 MIG/MAG Welding

MIG WELDING

Possibly the easiest welding technique to learn, good for the beginner.

Main applications

- Good for automotive panel work.
- Light fabrication.
- Can be used on stainless steels and aluminium with the correct filler wire and shielding gas.
- With bronze filler wire and argon gas, good for brazing galvanized and zinc-plated steel sheet without destroying the coating. Also good for modern automotive body panels containing boron steel.
- Using gasless filler wire, welding can be carried out in moderate winds.

The acronym MIG stands for Metal Inert Gas, referring to the type of shielding gas used to protect the weld pool while welding. An inert gas is one that does not react in any way with most other elements. In this case it refers to whether it reacts with the steel or whatever metal we are welding, since a MIG welder is able to weld aluminium and stainless steel with the selection of the right filler wire and shielding gas. Metal Active Gas (MAG), on the other hand, refers to the shielding gas being active on the metal that is being welded. To add further confusion, in North America MIG welding equipment is known as GMAW (Gas Metal Arc Welder).

Acronyms aside, let's now consider the uses of MIG welders, as we will refer to them from now on, whatever type of shielding gas is being used. We will later look at the increasingly popular gasless MIG

The MIG welder revolutionized welding, producing a clean finish without any of the slag characteristic of stick welding.

The inside of a MIG welding machine is more complicated, with moving parts and electronics, but this makes for relatively easy welding.

from the beginner's perspective, with an overview of what's available, what to use and, of course, when.

Inert Gases

You may have a vague recollection from school science lessons of a series of inert gases – helium, neon, argon, krypton, xenon and radon. On the periodic table these are referred to as noble gases, because under normal circumstances they rarely mix with other elements. This is due to the fact that at an atomic level the atoms of gas have a complete outer shell of electrons; if the outer shell of an element is incomplete then there is a tendency for that element to combine with another,

welders, which use a filler wire with a flux core so there is no need for a shielding gas.

As seen earlier, in order to create the right electrical conditions for conventional electrical welding equipment, a transformer of some kind is required to lower the voltage and thus increase the current available. The MIG welder is no different. It differs from an ordinary arc welder in that, instead of using welding electrodes to complete the electrical circuit, create the weld pool and add filler material, the MIG welder has a spool of filler wire that is fed into the weld pool, acting as the electrode. This is fed to the welding torch by a variable-speed motor through a flexible tube, which also carries the shielding gas in a separate pipe up to a hand-activated trigger. This switches on not only the feed motor, but also the main welding current, via relays, and a solenoid that releases the shielding gas into the shroud around the welding tip all the time the trigger is pulled.

SHIELDING GASES

A whole book could be written on the subject of shielding gases, but here we are looking at them

Shielding gas comes in various sizes and mixes for MIG welding.

CRYOGENIC GAS EXTRACTION

Most of the gases used in industry today are obtained by a cryogenic process. The air is cooled down until it liquefies. As the temperature is slowly increased, the various gases are drawn off at their individual boiling temperatures: first nitrogen, followed by argon, which boils at 87.3K, and then oxygen.

Kelvin (K) is a fundamental SI unit of measurement, named after the physicist William Thomson, Lord Kelvin, that defines the point at which all thermal motion ceases: 0K is known as absolute zero and is equivalent to $-273.15°C$.

forming a compound, as it tries to complete that shell by stealing electrons from other nearby atoms. This is why copper is such a good conductor of electricity, as electrons can flow from one atom to the next with ease, as its outer shell is incomplete.

The choice of argon rather than another inert gas is simply economics: argon happens to be the third most abundant gas in the atmosphere, after nitrogen and oxygen, at 0.93 per cent by volume. This is much more than carbon dioxide, which is much talked about but is only 0.039 per cent by volume in the atmosphere. Helium is used in some rarefied industrial shielding applications. Because it is lighter than air, however, it is not too good in a hand-held MIG torch, as it would try to go straight up into the air on leaving the torch: we want it to stay by the torch to act as a shield for the weld pool. Argon is ideal in this respect, as it is 25 per cent more dense than air; all the other inert gases are much less prevalent in the air and are more expensive to obtain.

Approximately 700,000 tonnes of argon are produced annually in the world by the cryogenic process. Argon also happens to be colourless, odourless and non-toxic, although it will act as an asphyxiant in enclosed spaces.

Active Gases

The most common active shielding gas is carbon dioxide (CO_2), which is used in welding mild steel. Although less prevalent than argon in the atmosphere, it is cheaper to produce as it is the by-product of many industrial processes; it can also be

distilled from the air, but this route is more expensive. It is not a pure element but a compound of carbon and oxygen, and breaks down readily in the heat of the arc to form carbon monoxide (CO) and free oxygen (O). Whereas oxygen usually pairs with another atom of oxygen to form O_2, the free oxygen will readily combine with other elements in the weld pool, leaving spatter and residue on the surface after welding. Benefits to the welding process outweigh these detrimental effects: it gives a hotter weld pool and increases the depth of penetration of the weld. This can be an advantage for thicker sections of steel to be welded.

Oxygen, although not used as a shielding gas alone, is often included in gas mixes for shielding purposes. Some stainless steels, in particular, require the addition of small amounts of oxygen, as it helps with the surface tension in the liquid weld pool.

Mixed Gases

It has been found that a mix of argon and carbon dioxide gives better economy than argon alone for general-purpose welding of mild steels, since carbon dioxide gives a hotter weld pool. The mixed shielding gases come with increasing amounts of carbon dioxide, from 5 per cent up to 20 per cent, depending on the thickness of material to be welded. Higher carbon content increases the depth of penetration, which is advantageous when thicker steel is being welded. The major shielding gas suppliers market their mixes under various trade names. The 95 per cent

A mix of 95 per cent argon and 5 per cent carbon dioxide is ideal for welding thinner steels and stainless steels with the MIG.

Pure argon is essential for aluminium welding and brazing with the MIG welder.

argon and 5 per cent carbon dioxide mix marketed by BOC, for example, is known as Argoshield Light; variants with increasing amounts of carbon dioxide in the mix are called Argoshield Universal and, with the most carbon dioxide for use on heavy steel sections, Argoshield Heavy.

SHIELDING GAS REGULATION

Whatever shielding gas is chosen, it will require some

form of pressure regulator to reduce the high pressure within the cylinder and give a constant flow of shielding gas as the welding proceeds.

The small disposable bottles use a screw-on regulator. Early ones came with a crude but effective flow gauge in the form of a ball bearing in a clear tube: the higher the ball in the tube, the higher the gas flow. The modern version has just a regulator knob and a gas outlet connection created by pushing the pipe from the welder into the connector, which self-locks; pulling back on a locking ring releases the

Small throwaway cylinders of shielding gas are available in all gases and mixes, but it is an expensive way to buy shielding gas. The cylinder of carbon dioxide shown here is suitable for welding steel.

useful guide as to how much gas is left in the cylinder, while another shows the amount of gas being output to the torch. The output given, although calibrated in litres per minute, is only an approximation of the flow of gas as it is actually a pressure gauge. Regulators are available with only a single gauge, but a twin gauge set-up can be purchased for just a few pounds more and is much more user-friendly. Once you are accustomed to your equipment, you will soon be aware from the sound of the gas coming through the torch, and the sound of the weld, as to whether you have too much gas or too little. Aim for a happy medium, since too much gas is not only expensive but is as bad as too little. If the shielding gas is forced

Throwaway cylinders have their own type of screw-on regulator.

pipe. Most DIY MIG welders come with this type of regulator, but if you are doing any more than the occasional repair weld, this is an extremely expensive way to buy your shielding gas.

All industrial-sized MIG welders use a regulator similar to those used in oxyacetylene welding. This is attached into the valve at the top of the shielding gas cylinder (right-hand thread) and the valve is opened before you commence welding. Regulators are available in various configurations. The best have two pressure gauges, one showing the pressure within the cylinder, which gives a

The industrial standard twin gauge regulator displays the cylinder contents remaining and the gas flow.

The alternative side-entry regulator on a hobby-type shielding gas cylinder.

through the welding torch shroud too quickly, it can induce eddies within the shroud and draw in air rather than excluding it, and this is detrimental to the weld pool. If there is too little gas flow, there will be insufficient to cover the weld pool, increasing the weld splatter and weld porosity.

Adapters are readily available to convert the pipe size on the DIY type welders with throwaway bottles to the standard industrial regulators. The cost of a few pounds will soon be recouped by savings in the quantity of gas used, if any amount of welding is being undertaken. Carbon dioxide cylinders used in industry have a different fitting to shielding gas cylinders; adapters are also available cheaply so that these cylinders can be used with the standard regulators.

For a modest fee a friendly landlord at your local pub might let you use a carbon dioxide bottle supplied for putting the fizz into beer and drinks. For several years I used such a bottle after a landlord I knew wanted the seat base in his car welded for the MOT test. It was a Sunday and my argon cylinder was empty. Since I didn't have the right adapter to connect my regulator to the carbon dioxide cylinder, I used a spare one from the pub. This gave me cheap welding until he gave up the pub, and of course it was a good excuse to sneak down there regularly.

WELDING PRACTICE

As before, rather than start by welding things together, try a few welds on a piece of plate to become accustomed to the welder and find the right settings for the material thickness and so on. It is a good confidence builder to know that the settings are right and that you can achieve the desired results before moving on to the real thing, such as repairing the wing of your classic car. There is nothing more demoralizing than to keep blowing holes where you are trying to achieve repairs.

SETTING UP THE MACHINE

Once out of the box, read all the instructions for your particular machine. Sometimes the machine will come with a small roll of filler wire installed. If it does not, check that the right roller for the wire feed is selected or is placed the correct way round, as often one roller will have two grooves for different-sized filler wires. The somewhat daunting task of initially threading the filler wire through the hose to the hand torch can begin. Install the reel of wire on the spindle with the correct tensioning springs. The end of the wire is usually anchored to the reel by being pushed through a hole in the rim and then

A pipe adapter is available to enable a DIY type of MIG welder to use the standard regulator.

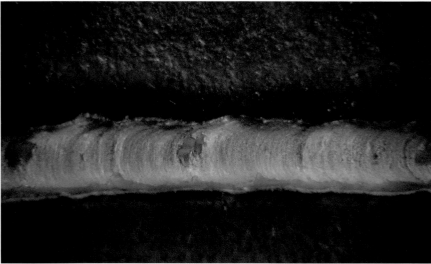

A smooth MIG weld needs very little finishing straight from the torch.

Try several welds at different machine settings, low through to high setting (top to bottom of picture). The top welds lack penetration; at the bottom they have too much! Somewhere between weld three and four will be fine.

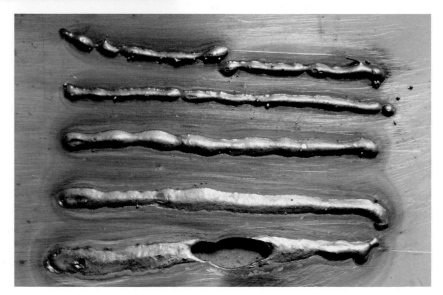

It is important to select the correct polarity when working with a dual use MIG welders. The control is usually at the front of the machine: gas shielding – torch negative; gasless – torch positive.

kinked over. Make sure that you have hold of the wire before cutting free with a pair of wire cutters, otherwise the free end will whip round and round with the released tension, unravelling many layers of wire and resulting in a tangled mess. This is not only wasteful of wire, but it takes time to sort out the mess.

Set the controls for the thickness of material that you are welding. The typical settings given in the instructions will be a good starting point to get you going, but you will soon learn to adjust the machine by the look of your test welds and the sound of the arc as you are welding.

If you have one of the dual machines that are capable of both gas and gasless MIG welding, make sure that the polarity to the torch is correct for the intended type of welding: the polarity of the torch tip for gasless welding is the reverse of that for standard MIG welding with gas. The usual arrangement to facilitate the changeover is by the machine's electrical connections, terminating at the front of the machine casing, instead of the wires being connected inside the machine, with the wires coming through, as in a single-use type machine. To complete the changeover, simply swap the two wires on the front of the machine marked with a '+' and the other a '[minus]'.

Pinch Rollers

Fundamental to consistent MIG welding is the correct setting of the pinch rollers. The usual set-up is to have a fixed roller driven from the wire speed motor, with an adjustable pinch roller with a groove around it, corresponding to the thickness of filler wire being used. It is important to select the correct-sized groove to fit the wire. The pinch roller is removable, held on with a screw or circlip. Many are reversible so that the different grooves line up with the wire position; this is critical, as the wire is fed up the end of the torch liner right next to the rollers. This roller is fitted to an adjustable spring-loaded carrier for adjusting the tension between the rollers.

If the roller pressure is too light, the rollers will slip on the filler wire, causing the filler wire at the torch tip to burn back into the tip nozzle, possibly welding the wire to the tip. Too great a pressure will not affect the welding, until something such as a poor earth connection stops the filler wire melting into the weld pool. The rollers cannot sense that the wire has stopped at the torch end and will keep pushing the filler wire up the liner. The usual consequence is that the filler wire kinks at the point of least resistance, a gap just in front of the rollers before the wire feeds up the liner. The rollers keep pushing and the resulting tangle of filler wire in this gap is often referred to as a 'bird's nest'. It is then necessary to remove the side of the machine to cut out the tangle before rethreading. The filler wire often jams when it reaches the nozzle tip, necessitating the removal of the tip. This can be screwed back in once the wire has passed out of the liner. This procedure is perhaps more annoying than anything else, but it all takes time in the middle of a welding job.

Selecting the correct setting for the roller pressure, before making any attempts at welding, is easy. Hold the torch nozzle about 25mm away from a hard surface, such as the workbench. Starting from slack, adjust the rollers until the wire is pushing against the bench with a force that is not enough to push the torch back towards you, but enough to feel and overcome the friction along the liner. Any more

pressure than this on the rollers will result in a bird's nest.

Nozzle Shroud Adjustment

Another fairly important setting is the relationship between the nozzle shroud and the nozzle tip. If welding by dip transfer, at low current settings, it is best to have the tip nozzle level with the shroud, or very slightly below. Any more than this will require the wire speed to be increased to compensate for the wire burning back, resulting in excess splatter being produced, and the bead form itself will suffer. When using spray transfer at higher currents, above 150 amps, it is advantageous to set the shroud about 6mm proud of the nozzle tip. This will reduce direct heat transfer from the hotter weld pool environment to the tip itself and help prevent splatter from entering the shroud.

Filler Wire Reel Tension

Most MIG welding machines are designed to use different-sized standard rolls of wire. The industrial standard is the 15kg roll, whereas machines intended for the DIY market have the facility to fit the smaller sizes of 5kg and the 800g mini rolls.

The pinch rollers are at the heart of MIG welding. The correct pressure is important for consistent welds. The roller shown has 0.6mm and 0.8mm grooves, with the latter serrated to grip the smooth, gasless MIG wire.

Industrial machines can also use the smaller sizes of reel. The bigger reel size is a much cheaper way of purchasing filler wire if you are using your welder a lot. If a specialized job comes up that needs a small amount of aluminium filler wire, for example, it would be wasteful to purchase a large reel as you would never use it all. The spindle on which the wire reel rotates has differing sized adapters for the different reel sizes. On the end of the spindle there is usually an adjustable spring-loaded arrangement intended to stop the wire reel from overrunning when the welding stops. Even a medium-sized reel of steel filler wire, weighing 5kg when new, will build up enough inertia to keep the reel turning after the feed rollers have stopped. The correct adjustment is critical to stop this overrun, but too much tension will make the feed rollers struggle to pull the wire from the reel. Even a small amount of overrun can cause the evenly wound rounds of filler wire on the reel to jump over one another. Any jerkiness in the wire feed will inevitably show in the finished welds, with at best an uneven finish. If the reel continues to turn, it is possible for a loop of filler wire to jump over the side of the reel. This creates more downtime as it will pull around the reel support shaft, which will need rectifying before the welding can resume. The worst-case scenario is that the looping wire creates a bird's nest on the reel side of the pinch rollers. This can be very wasteful on filler wire, as the usual remedy involves cutting out not only the bird's nest, but also many revolutions of wire to get back to the normal layers of wire. Once this has been sorted, the wire will then need feeding back up the torch liner again.

The correct tension of the wire reel is found when the reel stops as soon as the torch trigger is released. Any more than this will result in uneven welds.

DIP TRANSFER

MIG welding at low voltages and currents transfers the filler wire to the weld pool by what is known as 'dip transfer'. The welding wire dips into the weld pool and on contact immediately melts into the weld pool, burning back the filler wire. This process is

The correct nozzle shroud position is an important factor for good welds.

then repeated many times a second, producing the characteristic crackle associated with the MIG welding process.

SPRAY TRANSFER

At higher voltages enough power is available to melt the filler wire as it leaves the nozzle without dipping into the weld pool. This is known as 'spray transfer' and effectively sprays the weld pool with molten droplets of filler wire. Spray transfer produces more light and heat, and requires a powerful welder that can take filler wire thicker than the small 0.6mm diameter. It requires more than 160 amps to initiate this process.

In between the dip and spray transfer processes

Gasless MIG wire, with its flux core, is good for welding outside in draughts.

there is what is known as 'globular transfer'. This is to be avoided since, as the name suggests, the melted filler wire leaves the nozzle tip as large globules. Due to their size any magnetic effect from the arc has little effect on their direction and those globules that miss the weld pool end up as weld splatter. If using carbon dioxide as the shielding gas, however, you will not be able to avoid globular transfer, since a shielding gas with a high percentage of argon is required to create the right conditions for spray transfer.

GASLESS MIG WELDING

MIG welders have been developed that can weld without a shielding gas. Instead of an inert gas, a flux is carried as a core inside the welding wire itself. This is much like flux cored solder, so as the wire melts into the weld pool the flux is released and acts in a similar way to the flux-coated rods in stick welding, leaving a deposit to be removed once the weld is complete. One distinct advantage is that in an outside environment, where the vagaries of the wind come into play, the gasless MIG welder produces a better weld as the built-in flux does not disperse as readily as shielding gas from a bottle. Another benefit, especially for making repair welds outside on machinery too large to bring into the workshop, is that the welder can be carried without the need to take the shielding gas bottle and regulator everywhere. The downside is that, as the flux vaporizes in the heat of the arc, it produces a lot of whitish smoke, which somewhat obliterates the clear view of the welding process and has become synonymous with MIG welding. As with arc welding, this smoke contains many more detrimental compounds than when using an inert shielding gas.

When gasless MIG welding, the nozzle shroud can be removed to give a clearer view. This allows more space between the tip and work than is offered by shielded welding, with the result that the flux core heats up before entering the weld pool.

PLUG WELDING

Plug welding is a very useful technique when working with a MIG welder. It involves simply welding through a hole drilled or punched in the top piece where two are lapped. Its main use is in sheet work or joining car body panels that were originally spot-welded together. This can be useful

The weld produced with gasless MIG welding has a slag covering. After this has been removed, however, it is still not as tidy as a gas shielded weld.

when repairing vehicles to MOT standards, as the official stance is that structurally critical panels must be seam welded (continuous weld) or welded as manufactured. Where a repair panel is let into a sill structure, for example, the repair can be seam-welded around the periphery, but where it meets the flange at the bottom of the panel it can be spot or plug welded to the flange. This not only replicates the original look but it will be stronger than just welding it to the flange at the bottom edge.

Punched holes give a much cleaner hole than drilled ones, and holes made with a punch can be overlapped to give more of a slot for a larger and stronger plug weld. The two parts need to be clamped closely together at frequent intervals,

otherwise they will distort and separate as the panels heat up. Hold the torch nozzle perpendicular to the hole in the top sheet, at the same distance you would hold it for normal welding. Keep it still and pull the trigger, watching as the hole fills with weld metal. When it is slightly proud, release the trigger. At this stage it is to be hoped you have very little cleaning up to do. If the current was too low there will be an ugly build-up on the surface with a weak joint; conversely, too much current could have melted a hole in the bottom sheet before filling the punched hole. A slight increase in the usual current for the thickness being welded should suffice to produce a good weld, but always practise first on some scrap material to get the settings right before working on the real thing.

A hole needs to be punched through the top sheet in preparation for a plug weld, but the completed weld has a neat finish.

The underside of a plug weld showing full penetration.

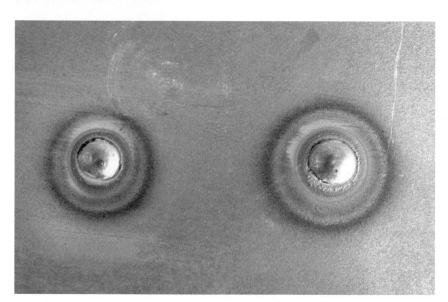

MIG WELDING STAINLESS STEEL

MIG welding stainless steel is much like welding mild steel. Stainless steel filler wirer is commonly available in two grades, 308L and 316L, both of which will weld the most commonly available 304 grade stainless steel. There is a possibility that the weld line might rust owing to heat from the welding displacing the chrome in the steel. The molybdenum content in 316L grade filler wire will counter the chrome displacement and gives relatively rust-free welds.

The linear expansion of stainless steel is about one and a half times that of mild steel, so distortion in thin material can be more of a problem. It is not insurmountable, however, with the use of tack welds and clamps.

The usual 95 per cent/5 per cent argon CO_2 shielding gas for stainless steel will usually suffice. Although the finished welds are perhaps not quite as neat as with mild steel, it will nevertheless produce serviceable welds.

The only extra thing to bear in mind when welding stainless steel is that, although fumes emanating from all types of welding are potentially dangerous to health, certain compounds of chrome, particularly hexavalent chromium (Cr(VI)), can be produced

An MIG welder can be used for stainless steel with the correct filler wire.

is much stiffer and there are fewer problems getting it to the weld pool.

Once the filler wire is changed, the shielding gas needs to be changed to pure argon. Since the aluminium filler wire is much softer than steel, it is advisable to change the lining inside the pipe to the torch. Teflon liners considerably reduce the friction between the liner and the filler wire as the wire is being pushed up the pipe to the torch nozzle. Unlike steel filler wires, which are much stiffer, the softer wire has a tendency to jam in the liner as friction is encountered. Professional MIG welding equipment specifically intended for welding aluminium and its alloys has the wire filler spool mounted at the welding end of the torch, thus eliminating or at least reducing the chances of the filler wire jamming up. Equipment with this option, of course, is more expensive than standard MIG outfits.

Preparation of aluminium for welding just needs a little care in what is used for the job. A stainless steel wire brush should be used to remove anything such as loose paint in order to eliminate the possibility of contamination from the iron bristles of a normal

in the heat of the weld pool and are known to be carcinogenic. Some stainless steels contain cadmium, which is also associated with cancer.

MIG WELDING ALUMINIUM

All that is necessary to weld aluminium with a MIG welder is to change the filler wire and shielding gas. The usual general-purpose grade of filler wire for use with aluminium MIG welding is 4043 grade, which contains 5 per cent silicon, lowering the melting temperature required in the weld pool. Problems can be encountered if a second run is necessary, as the metal in the line of the weld will have a lower melting point, owing to the alloying effect of the silicon deposited in the first run, and there is the real possibility of blowing holes.

A 5356 grade of aluminium wire is also available, but instead of silicon it contains magnesium, which gives the filler wire a higher melting temperature and makes it suitable for aluminiums with a magnesium content. The advantage with this grade is that the wire

Aluminium can be tackled with the MIG welder by using the correct filler wire and changing the shielding gas to pure argon.

The liner should be replaced if welding becomes jerky or rusty filler wire has been used in the machine.

wire brush. Special grinding and cutting discs are available for working aluminium, since the usual types have a tendency to clog with the relatively soft aluminium, quickly rendering them useless, and any discs previously used on steel will contaminate the surface.

Although the actual process of aluminium welding

with the MIG welder is the same as for welding steel, the base metal does not glow red hot as with steels, so it is harder to judge when a blow through is about to occur. It is imperative to push the weld with the torch, otherwise problems with shielding gas cover at the leading edge of the weld pool will occur.

MIG BRAZING

Once again, it is possible to braze with the MIG welder by changing the filler wire to a brazing wire made from brass and using pure argon for the shielding gas. This has several advantages to the budding MIG welder, increasing the range of jobs that can be achieved.

Developments in welding processes almost always come about from a need in industry to complete a job more easily and efficiently, or to weld together new materials that are being used for the first time. MIG brazing is no exception. Steels containing boron, for example, have been introduced by manufacturers striving to keep down the weight of new cars and increase the integral strength of the vehicle's safety structure. Boron contained within the steel lowers the melting temperature and using the MIG process, with a steel filler wire, to join boron steel has the effect of weakening it. Much lower temperatures

With the correct filler wire and pure argon shielding gas, successful welds can be made in aluminium.

By changing over to brazing filler wire and argon shielding gas, an MIG welder can be used for brazing.

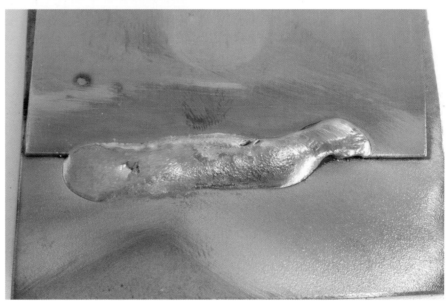

MIG brazing does not melt the parent metal and can be used on galvanized or zinc-plated sheet without damaging the coating or, as here, on stainless steel joined to mild steel.

are necessary when using brazing techniques, thus retaining the strength of the steel. Historically, brazing on a car's structure was always taboo but it is now recommended for structural repairs with boron steel. The equipment used specifically for this work is computer controlled and expensive, but an ordinary MIG welder will suffice for repair work with a little practice.

With this process it is possible to join galvanized and zinc-plated sheet without compromising the protective coating, join together dissimilar metals and repair castings made from cast iron.

MAINTENANCE

Regular maintenance of MIG welding equipment helps keep it working efficiently. All welding, apart from spot welding, produces some spatter to a greater or lesser degree. This depends on many variables, but some of the spatter is certain to find its way inside the gas shroud on the welding torch. If left, it will eventually block the holes through which the shielding gas flows, but before this happens the splatter will alter the gas flow characteristics, causing poor welds. Every now and again it is worth scraping

After a while the nozzle shroud and tip will be encrusted in weld splatter. Unless they are cleaned regularly, shielding gas flow will be restricted.

around inside the shroud with a small screwdriver or something similar to clear any accumulation. Once clear, it is a good idea to give it a squirt of anti-splatter spray. Another area requiring attention is the nozzle tip itself, which screws into the swan neck of the torch. It is from here that the filler wire picks up the welding current as it passes through the tip on its way to the weld pool. As with all electrical forms of welding, maintaining the electrical circuit is paramount for producing good welds. Keeping the resistance down by ensuring a good earth return is

Once the nozzle shroud and tip are clean again, welding can resume with confidence.

just as important at the business end of the torch. It is important to keep an eye on the tightness of the tip: the heating and cooling of the torch as it goes from the relative cool ambient temperature up to several hundred degrees or so, and back, has the effect of loosening the tip, thus increasing the resistance at this point. Not only will poor welding be a consequence, but this resistance may cause the tip to overheat to the extent that it melts into the filler wire. If the particular tip you are using has survived all of the above, have a look at the hole through it, as it will become worn as the filler wire passes through. A worn tip has the same effect as a loose tip, increasing the electrical resistance in the circuit. The tips must be seen as a consumable. You should keep several to hand, in the various sizes to match the wire used by your particular machine. They are not very expensive and are usually available in packs of five. This has the added advantage of keeping them clean in the packaging until required, as they will not be fit for purpose if left rolling around in a grubby toolbox for long.

At the other end of the torch, inside the machine itself, the rollers through which the filler wire passes also need routine maintenance. The pinch rollers themselves are usually made from hard material that does not wear too quickly. The usual steel filler wire used in MIG welding has a very thin copper coating that protects the wire from oxidation while on the reel. Although this is burnt off in the weld pool, a certain amount will turn to dust at the pinch rollers if there is any slippage of the wire. Any slight rust that forms on the filler wire on the reel will also rub off onto the rollers. You should occasionally remove the side cover from the machine and clean any accumulation from the rollers and the roller bearing surfaces, possibly lubricating where necessary. A blast through the inside of the reel housing with a blow gun fitted to an airline will shift any lingering dust. When changing reels of filler wire, a blast up the torch liner with the airline will move any wire dust that might cause binding within the liner at some stage.

Liner Change

The torch liner will possibly benefit from a change

Make sure you have a supply of spare nozzle tips to hand, as these are a consumable commodity when MIG welding. Check that you have the right size for the filler wire being used and the correct thread for your particular machine.

after a lot of welding. If the welding is becoming somewhat jerky or erratic, and other causes have been eliminated, a liner change will be beneficial. The usual steel coil or nylon liners are relatively cheap to replace, although replacement with a Teflon liner will reduce the friction substantially. If the equipment has not been used for a while, the filler wire on the reel inside the MIG machine will suffer from surface rust unless protected in some way. If the filler wire is used in this condition, loose rust will rub off inside the liner and cause jerkiness, even after replacing the filler wire with a new reel, until the liner is changed. To avoid the problem of rust on the filler wire, it may be a good idea to remove the reel and place it in a plastic bag, if not using the welder for some time, or simply cover the reel with a plastic bag, while it is still on the machine.

MIG WELDING PITFALLS

Easy to produce a weld that looks good, but weak through lack of penetration.

Small volumes of shielding gas are relatively expensive.

Large cylinder rental is expensive, unless undertaking a lot of welding.

Gasless MIG welding wire is expensive.

Inefficient when used in more than a light breeze.

The MIG welder has become the universal first choice of equipment in automotive repair work and light fabrication, as it will usually be sitting in the workshop ready to go with gas and filler wire installed.

The first technique to be shown is butt welding, which is common in the automotive repair industry. This is followed by a demonstration of making a brazed joint with the MIG welder, especially useful for modern steel panels containing high proportions of boron.

These examples are followed by a step-by-step guide on how to make a pair of heavy-duty shelf brackets for use in the workshop. No measurements are given as nothing here is critical: the most important thing is to get the angle as near to 90° as possible. The brackets were made from spare offcuts of 25mm mild steel angle iron.

A Butt Weld

1. After cleaning back to the bare metal, clamp the parts with the correct gap.

2. A coat of anti-spatter spray can be beneficial when cleaning up after welding.

3. Start up tack welding at the ends and avoid welding the clamps to whatever you are working on!

4. Tack at regular intervals across the panel.

5. Weld steadily between the tack welds, as though drawing a line with a pencil.

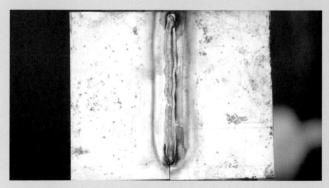

6. The finished butt weld.

Making a MIG Brazed Lap Joint

1. A clean and burr-free edge is required for a successful MIG brazed joint.

2. A thorough wire brushing ensures the surfaces are free of contamination.

3. For a lap joint to be successful, it is important that the two panels are clamped together tightly, as capillary action between them takes place.

4. After changing the filler wire, do not forget to use argon for the shielding gas. Tack at the ends and at regular intervals.

5. Once the panels are tacked together, run along the joint line at a steady pace in order to avoid too much build-up

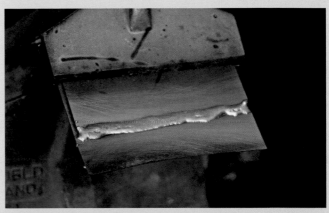

6. The finished job, leaving a neat joint. This technique is useful for joining galvanized or zinc-plated steel without damaging the coating. It is also effective on stainless steel or dissimilar metals.

How to Make a Pair of Heavy-Duty Shelf Brackets

1. Setting the correct gas flow is important for shielding the weld pool.

2. Cleaning off surface rust will help to give strong welds.

3. Marking out with ruler and scribe to ensure both brackets are the same size.

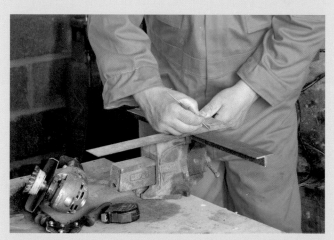

4. Using a square to mark across both pieces of angle iron. From this line mark 45° on both sides to give the 90° cut-out, allowing for the thickness of the steel.

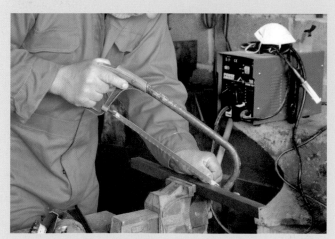

5. Cut along the marked lines with a hacksaw. This could also be done with an angle grinder.

6. If the marking out is accurate, a small, even gap will be left for weld penetration when the angle iron is bent at 90°.

7. Place a tack weld at the outside edge of the bend, not too large or it will be difficult to change the angle.

8. When the tack weld is done, use a square to check that the bend is at 90°. If the angle is incorrect and the tack weld is too strong to adjust, cut through it and start again.

9. Seam weld the angle using enough current to melt the edges together and penetrate the gap.

10. Cut two braces to strengthen the brackets, using lap welds at both ends. To keep the brackets looking neat, the first one was used as a template for the second.

11. Accurate and strong lap welds.

12. These may not be the prettiest brackets, but they are stronger than most available from DIY stores and it may have been quicker to make them than it would have been to go and purchase them. A coat of paint and they will be ready to attach to the wall.

6 TIG Welding

TIG WELDING

Excellent welds produced with practice. Greater dexterity required than with arc or MIG welding.

Main applications
- Good for joining thin sheet work in mild steel, stainless steel and aluminium alloys.
- Dedicated TIG machine with added foot control will give more control of current while welding, but requires practice.

The acronym TIG (Tungsten Inert Gas), although sounding similar to MIG, actually describes a wholly different process. TIG welding was until recently an exotic method, always out of reach and available only to professionals. The latest developments in electronic control, however, have made TIG welding affordable for the masses. The early development of techniques to join light alloys, especially in the emerging aircraft industry, has been described earlier (*see* Chapter One). The basic principles of using an electrical circuit to produce the heat for welding by this method are the same as for arc and MIG welding.

The main difference between the MIG and TIG processes is that, instead of using a filler wire/ consumable electrode to maintain the electrical circuit, the arc is made and maintained by a fixed but adjustable tungsten electrode. The holder for the tungsten electrode in the handset uses a collet system to grip the electrode and facilitate

Both hands are required for TIG welding, as well as a certain dexterity, not unlike gas welding.

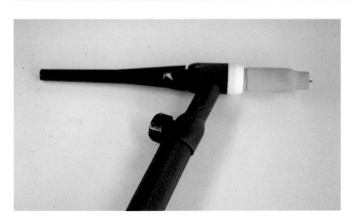

The heart of the operation, the TIG torch, containing the tungsten electrode.

negligible effect. The filler rod is added to the weld pool by hand, as in gas welding. This is the reason why TIG welding is deemed to be harder to learn than other welding disciplines. Owing to the extra manual dexterity required, and because the process has many more variables, TIG welding is much more

Once the technique is mastered, a TIG weld should be neat and tidy.

the adjustment. Different diameters of tungsten electrodes are used for the varying current levels for each size of electrode, each with a corresponding collet. Although the tungsten electrode erodes slowly during welding, and adjustment is necessary to maintain the distance by which it sticks out of the gas shroud, it plays no part in filling the weld pool. The very small amount of tungsten eroded will have

A typical add-on kit, comprising a torch with hose, cables and a regulator for use with an inverter welder.

versatile in the materials that can be joined together. Much finer work can be successfully undertaken because it is so much more controllable. The other outstanding feature is the quality of welds produced, which are unsurpassed other than by some exotic and expensive techniques.

In some respects TIG welding is similar to gas welding in that both hands are required along with similar hand-eye coordination: the heat from the torch is controlled by one hand, and the other is feeding in the filler rod. To the uninitiated, although a high level of dexterity is required, it is a bit like rubbing your stomach while patting your head at the same time. Having said that, it is not that difficult to master and many 'old hands' learnt to weld using gas equipment when it was the mainstay of many workshops.

A foot-operated pedal can be attached to the more sophisticated TIG welding machines to control the welding current available at the hand torch while welding is in progress. This is a very useful addition that offers the ability to reduce the current as you approach the end of a weld. At the end of a sheet, where the heat has nowhere left to dissipate, the risk of burning through is very real, so being able to reduce the current seamlessly is a boon. The other advantage is that the current can be reduced for thinner areas along the weld line that have been subject to corrosion – and should be cleaned before welding – again to avoid burn through. Conversely, it gives the operator the ability to give a burst of higher current to help the filler rod material flow at thicker sections. When welding aluminium and its alloys, an extra burst of current will also help lift a stubborn patch of oxide. This again requires further dexterity and additional expense, as foot-operated control pedals are rarely included when purchasing the TIG machine.

SHIELDING GAS

Most TIG welding uses pure argon as the shielding gas, although helium and other gases are used in some rarefied welding processes in industry. Although argon/carbon dioxide mixes are acceptable when MIG welding steel, if used on a TIG welding set-up the carbon dioxide content will have the

All normal TIG operations require pure argon as the shielding gas.

effect, once heated in the arc, of oxidizing the tungsten electrode. This is due to the carbon dioxide molecules breaking down into their constituent carbon and oxygen atoms when the electrode is at red heat. Tungsten has an affinity for oxygen at high temperature, similar to that of steel, and rapidly produces oxide compounds that will fall into the weld pool. This is detrimental to both the electrode and the weld quality.

SCRATCH START TIG WELDING

The most basic TIG welder is a scratch start DC machine. The arc is initiated by the tungsten electrode coming into contact with the item being welded; once the arc is initiated, welding can commence. This type of welder is fine for basic welding of mild steel and stainless steels. The downside of this simple machine is that it is not as versatile as the more expensive machines, and when struck, the arc is at the set current immediately, with the lack of any control to ramp it up.

LIFT START AND HF START TIG WELDING

The HF or high frequency start TIG welder eliminates the contact between the electrode and work surface required by the scratch start process. It superimposes a high frequency AC current over the main DC welding current, which has the effect of allowing a spark to jump between the electrode and the item to be welded, so as to initiate the arc

A foot-operated pedal can be used on more sophisticated machines, but it requires some skill to master its use.

before any contact between the two. The lack of contact is not that important with steel and stainless steel, but once moving on to aluminium it becomes imperative, as the weld pool will be contaminated, causing problems. The lift start function can be had without the high-frequency overlay on some DC only machines. Initiating the arc on scratch start machines is easier at a lower current, rather than starting at full welding current. The lift start raises the current, once the arc is initiated, to the full welding current, giving the operator time to focus on where they are going and prevent burn-through as the arc is initiated.

The dedicated TIG welding machine has become a sophisticated piece of equipment with a multitude of controls. This enables the operator to have much more control over the starting and finishing of the weld, as well as the welding itself.

The pre-weld shielding gas flow can be set along with the current ramp. This gives control as the arc is stuck, as the current starts at a low figure and goes up to the preset figure for the welding itself, with the ramp timing also being adjustable. On completion

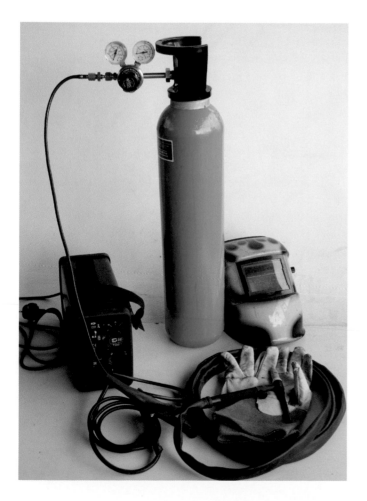

An MMA inverter welder set up for TIG welding, with scratch starting of the arc, is suitable for mild and stainless steel.

An AC/DC TIG welding machine is very flexible, but you should adjust only one knob at a time.

on the preset sequence at the start of welding and, on release, initiate the finish sequence. The switch can also be set as a latch, by which the sequence starts on pulling and releasing the trigger, while the finish sequence starts with a second pull of the trigger.

The benefits may seem a bit vague to the uninitiated, but it provides a competent operator with accurate control throughout the whole procedure, giving the TIG welding process a well-deserved reputation for producing the neatest and strongest welds. Top-end TIG welding equipment usually has the facility to add a foot control pedal, which not only takes over the trigger controls, but can be used to vary the current while welding is in progress: if the material being welded varies in thickness, then the current can be turned up or down to suit.

of the weld, the reverse is also fully adjustable as the current ramp can be reduced to finish the weld. The final welding current time can also be set on some machines, a feature known as crater control. If the relatively high welding current is cut quickly, then the weld pool has a tendency to freeze, resulting in a crater being left at the end of the welding run. This not only looks ugly, but has the potential to weaken the weld as cracking could occur from this point. With the ability to ramp down the current and with crater control, there is enough time for the operator to fill the crater as the current reduces. The shielding gas post-welding flow time is also adjustable, both to shield the weld until cooled sufficiently and to keep oxygen away from the tungsten electrode as it cools down.

All of these controls are set before welding commences, but this requires some way to control their sequences. The trigger switch on the hand torch is used to let the machine know when to initiate each sequence. This can be set to pull and hold, switching

TUNGSTEN ELECTRODES

The electrodes used in the early days of TIG welding were made from pure tungsten. It has been found that adding small amounts (usually 1 or 2 per cent) of some exotic-sounding materials – cerium, lanthium, thorium and zirconium, identified by a colour code – helps with striking the initial arc and maintaining the arc stability while welding. The other benefit of adding small amounts of these materials is that the current carrying capacity is increased for a given diameter of electrode.

Since a wide current range is available on the TIG welding machine for differing thicknesses of material, electrodes are available in a variety of sizes. For normal work, ranges are available from 1mm diameter, for thin material, through to 3.2mm diameter for thicker material. Thinner and thicker electrodes either side of this range are available for more specialist applications.

Tungsten Electrode Chart

Electrode Composition	Colour Code	Use	AC-DC
Tungsten	Green	Aluminium & Magnesium	AC
1 or 2 % Thoriated	Red	Mild & Stainless Steel	DC
1% Lanthanated	Black		
Ceriated	Grey	Steel & Aluminium	DC-AC
1% Zirconated	White	Aluminium	AC

Colour code shown for UK and Europe.

Tungsten Diameter	Plate Thickness	DCEN*	DCEP**	AC
0.5 mm	Under 1mm	5–20		5–20
1.0 mm	2 mm	15–80		15–80
1.6 mm	2–3 mm	70–150	10–20	70–150
2.4 mm	4–5 mm	150–250	15–30	110–180
3.2 mm	5 mm	250–400	25–40	150–200

* DC current electrode negative
** DC current electrode positive

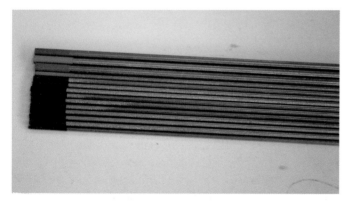

Tungsten electrodes come in a variety of sizes for differing currents.

Tungsten electrodes may be identified by a colour marking on one end: here are red thoriated rods for DC use and grey ceriated rods for AC or DC use.

GRINDING THE POINT

Although electrodes sometimes come with a pre-ground tip, the tip will require forming once some welding has been carried out or if the electrodes come with a square end. A bench grinder fitted with a fine aluminium oxide wheel will suffice for limited amounts of regrinding. Grinding lines must be parallel to the length of the electrode; this has the effect of concentrating the heat at the tip. Grinding the tip with the lines concentrically with the electrode diameter will lead to an unstable arc and uneven heating of the weld pool. For low current work, an angle of about 30° will be fine,

To produce and maintain the arc with DC TIG welding, the tip needs to be ground to a point with the grinding lines running parallel with the electrode. Dust from thoriated electrodes is potentially harmful, so always wear a face mask and observe safety instructions.

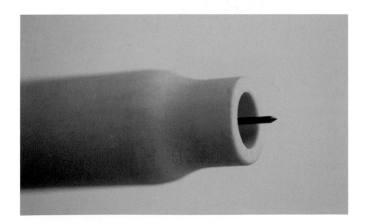

The distance the tip sticks out of the shroud is a compromise between your ability to see what is happening and the gas shielding required: 3–6mm is about right.

but for higher currents up to 100 amps, angles of 60, 90 or even 120° may be necessary. For anything but very low currents, then, the tip should be ground slightly flat to take off the sharp point. This will keep the tip in shape for longer, but bear in mind that the more the electrode is ground, the quicker it will be used up. For AC welding the tip shape does not affect the arc as much as with DC. Here just a chamfer on the square end will suffice, since the tip will become rounded as work progresses. When DC work is being undertaken, it is the tip shape and condition that determines whether a stable arc will be sustained, and careful redressing of the electrode tip now and again will pay dividends.

HEALTH WARNING

Although the risk is small, it must be mentioned that thorium is a low-level radioactive source, emitting alpha particles. General handling of electrodes containing thorium poses minimal risk to health, provided they aren't handled with open cuts on the operator's hands. The biggest risk comes from contaminated dust entering the lungs. The Health and Safety Executive has stated that, if possible, an alternative non-thoriated electrode should be used, or the electrodes should be reshaped with a purpose-built grinder that captures the dust produced. A good breathing mask should be worn while grinding for the occasional reshaping. The dust is damped and collected into a sealable plastic container, and disposed of at an approved landfill site.

TIG WELDING ATTACHMENTS

Most DC inverter welders designed for mmA arc welding may be adapted to take a kit that can carry out rudimentary TIG welding. These lack the finer controls available on a fully fledged TIG machine, such as rising start current, but can be very useful for welding thin-section mild steel and stainless steel. The kits invariably contain a TIG hand torch and shielding gas regulator. Since the leads on inverter welders are of the plug-in

Once the arc is established, it is bright and smooth, unlike the MIG welder's crackle.

The regulator set-up is the same as on an MIG welder. Twin gauges, showing cylinder contents and gas flow, are best.

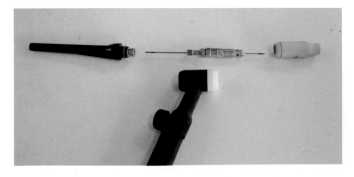

Exploded view of hand torch showing components.

type, it is easy to change from electrode positive to electrode negative, which is the usual choice, although electrode positive can be used in certain circumstances.

GETTING STARTED

So much for the theory, now let's get practical, starting with an introduction to basic scratch start TIG operation. The versatility of the TIG process will not be fully realized until after using one of the more sophisticated pieces of equipment, but initially the array of differing controls on these machines will be confusing and of little help to the uninitiated without a grasp of the basics.

The add-on attachment to an inverter mmA welder is a good place to start as an illustration of the set-up procedures. The basic kit consists of a TIG hand torch with cable and gas pipe attached, together with the earth lead and clamp from the inverter welder. The kit may or may not contain a regulator for the shielding gas. As already discussed, this will be argon, to regulate the pressure in the gas cylinder to a suitably low level. If MIG equipment is already being used, it will be possible to 'borrow' its regulator. It will make sense, however, to purchase a new one so that there isn't a mad scrabble around the workshop looking for the regulator whenever differing welding disciplines are undertaken.

The regulator set-up is fairly standard and has been covered elsewhere, so attention will now turn to the business end, the hand torch. The usual kits available come with one tungsten electrode (probably 1.6mm diameter, 2 per cent thoriated) suitable for use on relatively thin mild and stainless steels, the materials that are the mainstay of what can be welded with a basic DC machine. The tungsten electrode will require a point ground on the working end. Either end could be ground, but it is good practice to grind the unmarked end as, if you have various types of electrode in stock, there is no way to identify the type of electrode after the coloured marker is gone. The electrode needs to be sticking out about 3–6mm proud of the ceramic gas shroud. This is then locked in position by tightening the back cap, which pushes down on the collet that holds its position. As the electrode erodes, the back cap may be unscrewed to release the electrode for adjustment or for regrinding the point. The torch lead should be plugged into the negative socket on the inverter; this is usually locked with a quarter turn, as the plug has a boss sticking out from the side, and the socket has a bayonet arrangement inside. The earth lead is attached in a similar fashion to the positive socket. Once the gas is turned on at the cylinder, the flow can be adjusted at the regulator with the control valve on the handset turned on. The flow rate should be somewhere between 4 and 10 litres per minute. This can be fine-tuned as you get to grips with the practicalities of the process, since too much shielding gas does not improve the weld, it just wastes expensive gas.

All that remains is to set the current control on the inverter. Some have an mmA/TIG switch that controls the open circuit voltage (OCV) of the machine. When butt welding mild steel of car body thickness (approx. 0.8mm), for example, a current of around 20–30 amps will be about right, but again can be fine-tuned as you practise.

With TIG welding, the watchword is comfort. Get yourself in a relaxed position, do not forget to put the earth clamp on, and you are ready to start. It is a good idea to have some means of relieving the weight of the hose to the hand torch, so that gravity is not trying to pull the torch away from the weld area. Do not forget to don gloves and eye shield, although a self-darkening helmet makes sense for easier handling. Turn the gas flow on at the hand torch and gently scratch the electrode on the plate to be welded; a light touch is required, as the electrode may stick until you are practised. As you see the initial spark, raise the torch by 3–5mm and the arc should initiate in the argon stream. Once the arc is established, tilt the torch over to an angle of

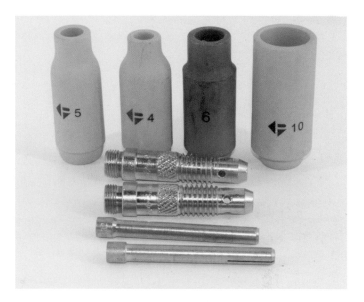

Collets and collet bodies are available for each size of electrode.

60–70 degrees, pointing in the direction of welding as with MIG and arc welding. Bring a suitable filler rod in at a low angle, keeping the end in the stream of shielding gas, so as to limit oxidation of the hot rod. Practising on plate before actually welding a joint will help to show the characteristics of the metal being welded, at the same time as you gain confidence. If you start on a joint, the usual result is that there is so much to take in at once that you miss

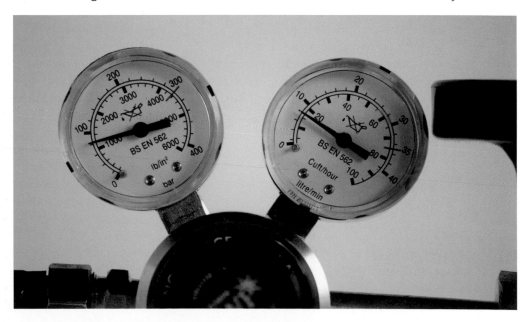

Shielding gas should be regulated to between 4 and 10 litres per minute.

For steel welding, the hand torch should be held at about 60–70°, with the filler rod at about 30°.

seeing one side of the joint melting away. Once you are ready to make a welded joint, start by tacking at the ends and at regular intervals. When tack welding, a circular motion once the arc is initiated will help to flow the two pieces together as you dab in the filler rod. Once tacked together the joint can now be welded, adding filler rod to the weld pool as required and, if done well, leaving the signature smooth, evenly rippled finish of a TIG weld. Once the weld is complete, leave the gas flowing until the tungsten electrode has cooled, since the red-hot electrode will oxidize in the presence of air if it is shut off too quickly.

OBSERVATIONS

After many years using MIG and gas welding, several observations come to the fore. The most noticeable

feature of the simple scratch start TIG is that the shielding gas control is manual, whereas with MIG welding the gas flow is under control automatically on pulling the trigger and is shut off on releasing it. With TIG, however, the gas has to be turned on before welding and turned off after the weld is complete. The main differences with gas welding are twofold: the initial heating of the weld pool is much quicker with TIG welding and the welding position of the operator is much more critical as the gas shroud partially blocks the view of the proceedings, whereas with gas welding you have a broader view of the surrounding area. The most obvious advantage is the relatively small amount of distortion with TIG.

If you have any previous experience of mild steel welding, before embarking on anything more exotic with the TIG welder, such as stainless steel, practise on some mild steel. This way you can draw on past experience of how the material behaves under other welding disciplines, so that the only variable is the welding heat source. Once you have mastered the correct technique on the material you are familiar with, then move on to a different material. This will instil confidence in your ability, as you will know that you can produce good welds with equipment new to you.

Earlier it was mentioned that the simple scratch start TIG welding attachment does not have any controls. Once you have grasped and put into practice a basic understanding of the principles involved, it will be much easier to move on to a more sophisticated machine with a multitude of controls, which can then be adjusted one at a time from a position of knowledge, rather than stabbing in the dark.

One of the advantages of the TIG process is that filler rods are rarely required for thin section and outside corner welds as the edges melt together, leaving a very neat finish to the welds. To be successful the edges need to touch all along the joint line, otherwise the edge will melt away on meeting a gap, rather than melting together.

AC TIG WELDING

Once you have practised welding various steels, moving on to aluminium and its alloys will present new challenges. The thoriated (red tip) electrode used for DC work will need changing to either a general-purpose ceriated (grey tip) electrode, which

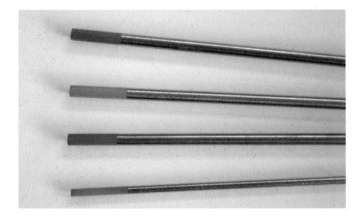

General-purpose ceriated electrodes, which are designed for use with both AC and DC, can be used on steel as well as aluminium.

The correct preparation of aluminium is essential. In order to avoid contamination, use a stainless steel wire brush and zirconium flap wheel.

can be used on both AC and DC, or a zirconated (white tip) electrode specifically for use with AC current. The electrode does not need shaping to a point as with DC, since with AC current the end of the electrode will maintain a ball shape. If a large electrode is being used, then bevelling the edge will be beneficial to initiating the arc. The biggest problem with aluminium, especially if repair work is being undertaken, is identifying the actual composition of the alloy used; if new material is purchased, the specification will be available from the supplier. Once the machine is set correctly, including switching it to AC and HF start, it is imperative that the electrode tip does not make contact with the surface being welded, as this will contaminate the electrode before any welding has taken place. Bring the electrode to within 3mm of the surface and start the TIG machine sequence by pressing the trigger switch. The arc will strike up as the HF burst ionizes the gas already emanating from the shroud, allowing the arc to establish itself as the current comes up to the selected value for the work in hand. Whereas the arc is almost silent when using DC current, a loud hum comes from the arc with AC. It appears much brighter and is more of a green colour, rather than the blue of DC. Now will be the time to tell whether the aluminium has been properly cleaned to remove any dirt or the thick layer of oxide that can form. It should first be wire brushed with a stainless steel brush and then, for good measure, a zirconium flap wheel should be

When welding aluminium, the torch needs to be more upright in order to prevent the filler rod melting before it gets to the weld pool.

run along the edges to be joined. You should also rub the filler wire to help remove any oxides. While welding aluminium, it will be found easier to keep the torch more upright and bring the filler rod in at a low angle, keeping the hot end of the filler rod within the shielding gas envelope so as to prevent it oxidizing. If the usual torch angle is used, the filler rod will melt prematurely. Higher currents than with steel will need to be used for the same thicknesses of aluminium: although it melts at a much lower temperature, it is a far better conductor of heat. The hardest job is to initiate the weld pool as the two edges melt, but once established, rapid progress can be made without further problems. The foot pedal is a very useful addition to the TIG welder's armoury, especially when working with aluminium, as that extra burst of current applied will make a difference to the success of the weld if oxides are encountered during welding.

GAS SHROUDS AND LENSES

As mentioned above, cranking up the gas flow does not necessarily improve the shielding of the weld pool. Indeed it will possibly be detrimental as it will cause swirling in the shielding gas, which will possibly draw in the surrounding air. This will become apparent from the condition of the arc. Shrouds of differing diameters are readily available for all the standard TIG torches and it is essential to choose the right size for the electrode used: the loose formula is that the shroud number selected should be approximately four times the electrode diameter (for a 1.6mm diameter electrode, for example, the shroud should be a No. 6).

One of the disadvantages as the shroud size becomes bigger is that it obscures the operator's view. It might appear that you only need to move

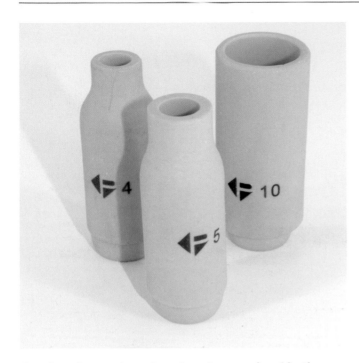

The gas lens focuses the shielding gas into a parallel column, giving better coverage and using less gas.

Gas shrouds come in various sizes. As a rough guide, the shroud number should be four times the electrode diameter.

On a standard collet body the gas issues from annular holes, but with a gas lens it comes through a fine mesh in the end.

the electrode so that more is sticking out of the shroud, but when working on the flat any more than 6mm will be detrimental, and just 10mm is possible when working on a fillet weld where the shielding gas is partially restricted.

A gas lens can be used to overcome this problem. The standard collet body is replaced with the gas lens, the design of which creates a parallel stream of gas. The standard collet body has four holes around the periphery at the shroud end that allow the shielding gas to enter the shroud, but in no particular order. The gas lens has a very fine mesh through which the gas passes at the shroud end of the collet body; this has the effect of focusing the shielding gas stream, hence the name 'gas lens', into a parallel column that removes any turbulence. It is now possible for the electrode to stick out further, giving a better view of the weld pool, and the TIG welding is less affected by draughts. Despite these positive features, there is also a downside: the gas lens is more expensive than a standard collet and one is required for every size of electrode used.

The gas lens requires a larger body on the gas shroud to accommodate it, but allows the electrode to stick out further.

TIG WELDING PITFALLS

Greater manual dexterity required than with arc or MIG welding.

Argon shielding gas expensive.

AC TIG machines expensive to purchase for use on aluminium.

Dangerous fumes produced when welding stainless steel.

The light from TIG welding is extremely bright.

The standard shroud also needs to be replaced with one that matches the larger diameter of the gas lens. Although they cost more, the use of gas lenses should see a saving on shielding gas as the unified stream permits a reduction in the gas flow.

TORCH BODIES

The TIG process is a relatively recent introduction from industry. Hand torches are made in standard sizes, so that if you have a WP 20 or WP 26, for example, then all accessories for that particular torch will fit. To extend the flexibility, the standard fixed-head hand torch can be replaced with one with a flexible neck, which gives a better approach to difficult positions and makes the welding much easier. If you have a WP 20 standard hand torch, for example, purchase a flexible head WP 20 and you will find that all your collets and gas shrouds fit without having to purchase a whole new assortment. On the subject of torch bodies, the back cap on the torch is long enough to accommodate the 150mm electrodes. In certain circumstances, however, such as when repairing machinery, this long back cap may perhaps be in the way and catch on things. A short back cap can be purchased to alleviate this problem, but unless a part-used electrode is available an electrode will need to be shortened to fit. The better quality electrodes will require a nick cut with a grinder to allow it to snap at the required length, although the earlier warnings about thoriated electrodes should be kept in mind.

The TIG welder is now in easy reach of the serious DIY welder. The introduction of the DC inverter welder and the ability to attach a TIG torch means that it is easy to get started. The first exercise will be to put a stainless steel bracket on a stainless steel exhaust pipe, followed by the joining of a mild steel exhaust system. The latter could just as easily have been completed with gas welding or MIG welding, but here the TIG welder was used to demonstrate the control possible on thin materials.

Attaching a Bracket to a Stainless Steel Exhaust Pipe

1. An angle grinder fitted with a zirconium flap wheel, essential for cleaning stainless steel to avoid contamination, lies ready with the TIG torch before welding a bracket on a stainless steel exhaust pipe.

2. Ensure the gas flow from the argon bottle is between 4 and 10 litres per minute.

3. Clean the exhaust pipe and bracket thoroughly with the zirconium flap wheel.

4. Use a clamp to reach around the pipe, leaving enough room for the TIG torch, and clamp the bracket into position.

5. After finding there were no stainless steel welding rods in stock at the local welding supply shop, I cut a thin strip from the same sheet from which I made the bracket. More heat was concentrated on the bracket since it is thicker than the pipe.

6. After the clamp is removed the other side is welded. Cleaning with a wire brush leaves a neat weld.

Joining a Mild Steel Exhaust Pipe to a Silencer

1. Looking for a silencer to replace the one on my old Land Rover, I found a serviceable one in the scrap bin and a length of pipe of the right shape.

2. The TIG welder was used as it gives easily controllable welds on thin metals.

3. The parts were thoroughly cleaned with a wire wheel before they were aligned and tack welded.

4. On close inspection the tack weld had some blow holes, possibly from rust contaminating the weld pool.

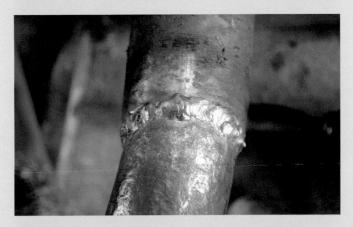

5. Working around the joint until the weld is complete is effectively creating a lap joint.

6. The finished job ready for fitting and hopefully an MOT pass.

7 Gas Welding

GAS WELDING

Requires higher levels of dexterity than with most techniques.

Main applications
- Good for sheet metal work, brazing of steel and aluminium alloys.
- Cast iron repairs possible with the correct filler rods.
- When used with the appropriate nozzles, excellent for cutting steels and heating applications.

Gas welding was once the mainstay of body shops around the world, but has been largely superseded by MIG welding and, more recently, TIG welding, both of which are quicker and produce less distortion in thin panel work. The beauty of using gas welding equipment is its versatility in the right hands, ranging from the smallest flame produced from a No. 1 nozzle, fitted in a 'model O' torch, through to a flame with copious amounts of heat from a No. 25 nozzle, or bigger, in a standard torch such as a Sapphire 5. Although it may perhaps lay beyond the remit of this book, fitted with a heavy-duty mixing chamber and a pepper-pot nozzle, so-called because it has several holes around the periphery, it can be used for such tasks as heating seized components or items that require bending. When a dedicated cutting torch or cutting head is fitted to the standard torch body, gas equipment is very efficient at cutting steel and sometimes known colloquially as the gas axe, owing to its cutting speed. Nozzles of differing sizes are available to tackle thin sheets up to slabs 12in (300mm) thick.

Gas welding can be spectacular at times, with copious showers of sparks.

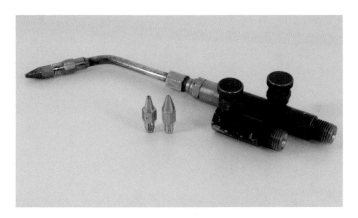

Gas welding gives versatility at your fingertips. The model 'O' welding torch, shown here, produces a very small flame but at a high temperature.

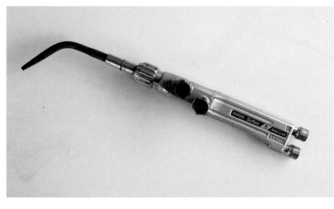

The Saffire 5 standard gas welding torch, shown with a swaged welding nozzle, can be used for welding, cutting or heating with the right attachments fitted.

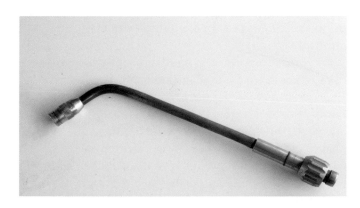

The pepper-pot oxyacetylene heating nozzle is named after its ring of holes.

ACETYLENE GAS

Acetylene is an alkyne with a chemical formula of C_2H_2. Each molecule consists of two carbon atoms and two hydrogen atoms, held together with a triple bond. Without wishing to get bogged down in chemistry, it is enough to say that this results in a very unstable gas. Today it is mainly produced by the partial combustion of methane, but it can be produced by organic means. When in contact with silver or copper, it readily produces acetylide compounds, which are extremely explosive. Burnt with oxygen, acetylene produces a flame with a temperature of about 3,500°C, hot enough to weld steel. Acetylene is unique in that it can burn in air from as little as 2.5 per cent right up to 80 per cent.

Later in this chapter there will be discussion of setting up and using cutting gear with gas equipment, since a cutting head is usually included if you buy a complete kit to start gas welding.

A distinct advantage with gas welding equipment is its portability. Because it is self-contained, especially when the smaller sizes of cylinders are chosen, it can be transported to places where no electricity supply is available. There is the usual trade off: large cylinders – cheaper gas, smaller cylinders – more expensive gas. To offset this, though, smaller cylinders are cheaper to hire.

Oxygen and shielding gas fittings have normal right-hand threads and plain nuts.

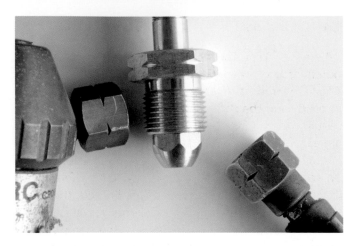

All fuel fittings have left-hand threads, indicated by a notch on each corner of the fitting nut.

Oxygen and shielding gas fittings have normal right-hand threads and plain nuts.

THE OXYACETYLENE FLAME

When acetylene is burned with oxygen at a ratio of 1:1 at the tip of the welding torch, it produces a bright blue inner cone. This is the primary combustion. The secondary stage produces the paler blue envelope, created by combining about one-and-a-half parts of oxygen, from the air around the torch, with carbon monoxide and un-burnt hydrocarbons from the primary stage, giving off more heat. For complete combustion of every part of the acetylene, two-and-a-half parts of oxygen are required.

The only other gas available that produces enough heat for welding steel is MAPP gas, a combination of methylacetylene and propane. This gas can be compressed without the dangers associated with acetylene and is usually sold to the plumbing trade in small throwaway canisters. The temperature produced when this fuel gas is burnt with oxygen is only slightly lower than with acetylene.

SETTING UP THE EQUIPMENT

Equipment for gas welding has evolved over the years and is fairly straightforward. Unlike electric welding, where it is only possible to electrocute yourself, if things go wrong with gas equipment it might take out the whole neighbourhood! There should be no problems, however, if you adhere to a few stringent safety rules.

Safety First

The standards that have evolved include that all flammable gases, such as acetylene and propane, use pipefittings and cylinders with left-hand threads, marked by a notch cut on each corner of the nuts. All non-flammable gases, such as oxygen, argon and carbon dioxide (the last two of which are used as shielding gases), have right-hand threads. It may be asked why oxygen is down as a non-flammable gas. It does not burn on its own if you try to light it, but it will support and enhance combustion. Without oxygen normal combustion will not take place: when a gas burns in the air, it is using the 20 per cent of oxygen contained in the atmosphere.

It is vitally important to heed a number of simple rules when using or handling gas cylinders (*see* Chapter Three). Never use gas cylinders for rollers. Never drop or hit gas cylinders together or with anything. When in use, always secure them in a proper cylinder trolley or chained to the wall of the workshop, since it may well be a disaster if a cylinder were to fall over. Oil and grease will spontaneously combust in the presence of oxygen, so never use oil or grease on any gas welding fittings, and never drape overalls over the cylinders and regulators. Never under any circumstances be tempted to use oxygen as a substitute for compressed air.

Safety Devices

From the very beginning, safety features have been incorporated as oxyacetylene equipment evolved in an attempt to eliminate any possibility of a flashback from the welding or cutting torch. The danger with a flashback is the phenomenal speed at which it travels back along the hoses, creating a shock wave that could induce self-combustion inside the acetylene cylinder. The first line of defence back from the torch is the one-way check valve, which lets the gas through but should shut with the back pressure during a backfire in the torch. If this fails to arrest the burning gas in the hose, the next line of defence is the flashback arrestor, which must be fitted between the hoses and the pressure regulators. The flashback arrestor is designed to cool the flame temperature to below the ignition temperature of the gas, extinguishing the flame. It is backed up by a cut-off valve that is either reset automatically when cooled or is of the resettable flashback arrestor type, which is manually reset after being activated by a flashback or, since they are so sensitive to back pressure, if the pressure in the hose is higher than in the regulator.

Gas Cylinders

Whereas welding torch, hoses and regulators are purchased outright, gas cylinders are another matter. They are hired from large companies that specialize in producing the various gases and their bottling requirements. Once the gas is used, the cylinders are exchanged for a full one, just paying for the gas and a handling charge. The cylinder rental itself is on an annual, three- or five-year payment scheme. If your gas welding requirements are limited, then the small cylinders will suffice. Cylinder rental is less for these small sizes, but the gas is proportionally more expensive. Conversely, the bigger cylinders have a higher rental charge, but the gas works out cheaper. The cylinders themselves obviously increase in weight as they get bigger. This may seem like restrictive practices but, from the safety point of view, the cylinders and valves are given safety checks each time they are returned. The normal procedure is to open an account and then you can hire cylinders on a monthly basis plus the gas, returning the cylinders when they are finished. Since most of these companies have branches all over the country, there is the advantage that you can go into any branch if you are working away from home and get a refill on your account. Most of them stock any filler rods and other equipment you might need, but these can be

An acetylene cylinder is maroon in colour and has the test pressures to which it is tested stamped around the neck.

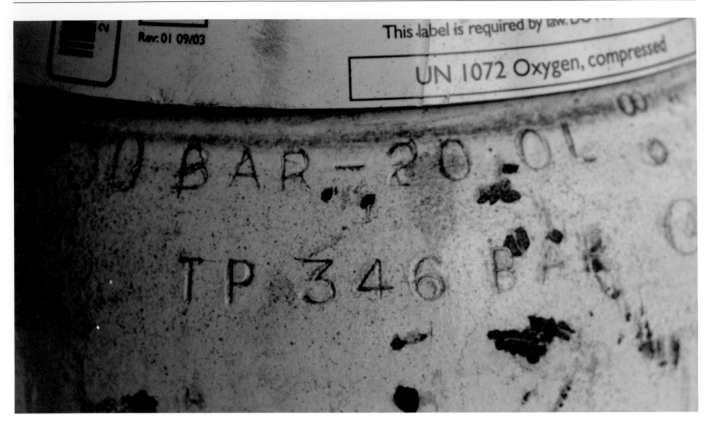

An oxygen cylinder, painted black with a white top ring, is also stamped with test and service pressures.

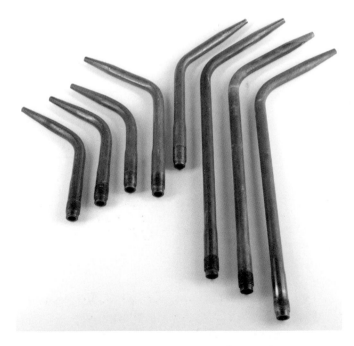

Swaged gas welding nozzles suitable for most welding needs: (left to right) Nos. 1, 2, 3, 5, 7, 10, 13 and 18.

expensive. The staff behind the counter, though, are trained in the use of their products and can be most helpful to the novice gas welder.

Putting It All Together

The gas cylinders, whatever size you are using, should first be secured into a suitable trolley or chained back against the workshop wall. After removing the plastic caps fitted by the gas supplier to protect the valve threads and keep out dust, the cylinders need to be 'cracked'. This is done by momentarily opening the valve so that the release of pressure blows away from the valve seat any debris or water that might have found its way there and would stop the regulator from seating properly, possibly causing a leak of gas. While doing this it is imperative to look away or wear safety glasses, as any foreign bodies lurking in the valves will become missiles once the pressure is released.

start in a tangle they will be in a tangle for the duration of the job and will be a potential trip hazard. Both ends of each hose will possibly have a fitting of the same size, but one end will have a check valve fitted. This must be fitted to the torch, as it is a one-way valve that lets the gases into the torch but snaps shut if there is a backfire, preventing it from travelling up the hoses to the cylinders. Tighten all fittings gently but firmly with the correctly sized spanner. Do not overtighten.

The valve should be cracked open momentarily to blow out any foreign bodies from the cylinder valve. Always avert your eyes and wear goggles.

When fitting the regulators to the cylinders, always use the correct spanner, and remember that oxygen has a right-hand thread and acetylene a left-hand thread.

The regulators should next be connected, remembering that connectors for flammable gases have a left-hand thread and turn anti-clockwise when seen from above. These are followed by the flashback arrestors. Once these are fitted, make sure that you can reach the valve knobs on the cylinders, if these are fitted, or that you can easily reach the valve with the key. Slacken the regulator nut and change the position of the regulator in relation to the top of the cylinder so that it gives better access to the valve. The hoses are next and follow a standard colour code, with blue for oxygen and red for acetylene in the UK (it should be remembered, though, that in the USA it is green for oxygen and red for acetylene). At this point it is worth running out the hoses in order to remove any coils in the pipe before connecting them to the torch; if they

Before turning on the gas valves for the first time, check that the regulator adjusters are fully wound out. They should always be left in this state when you have finished, as it unloads the springs inside the regulator, maintaining accuracy for longer. Turn on

*Hoses (red for acetylene, blue for oxygen)
should be rolled up when not in use.*

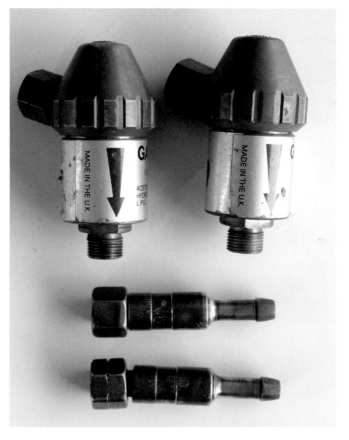

ABOVE: *Safe use of gas equipment requires flashback arrestors
at the regulators and check valves fitted at the torch end of
the hoses.*

LEFT: *If hoses are left like this, it won't be long before someone
trips over them.*

Once all is connected, the cylinder valve can be opened. The left-hand gauge shows the oxygen cylinder pressure and the right-hand the pressure to the torch, in this case 40–50 psi for cutting.

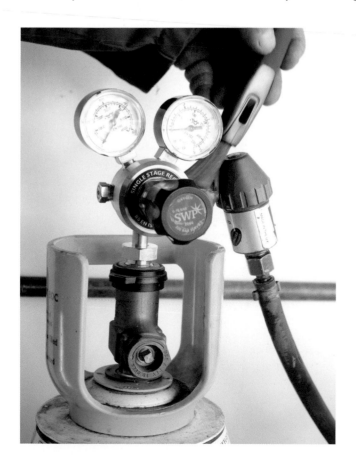

the gas valves at the cylinders, one at time, starting with the oxygen. The high-pressure gauge will now be registering the pressure within the cylinder and can be used as a rough guide as to how much gas is left in the cylinder. Turn the regulators up until the gauge registers $2lb/in^2$, which is sufficient for nozzles up to a No 7. Open the valves at the torch for a few seconds in turn to purge the hoses of air. While doing this, readjust the regulator to maintain the stated pressure. At this stage it is prudent to test the whole outfit for any gas leaks. This can be done with a can of proprietary leak spray, which is sprayed on all the joints in the set-up. Any bubbles issuing will indicate a leak from that joint. A cheaper alternative is to apply soapy water with a brush. Any leaking joints can then be further tightened. If this does not effect a remedy, undo that joint and clean the connection seats, before remaking the joint and retesting.

A proprietary joint-testing solution or soapy water is painted on to the joints to check for gas leaks.

Bubbles issuing from any joint indicate a leak of gas. Check again after retightening the joint.

Lighting Up Procedure

It is always best to start as you mean to go on: only use a proper spark lighter to light the gas. It is tempting to use a disposable gas lighter – cheap, convenient and available for most purposes – but remove it from your pocket before venturing out of the house. It is not the fact that it will not light the gas: it is the potential danger that the red-hot end of a welding rod might go through your overalls and into your leg. This will be painful enough, but if it penetrates a gas lighter in your trouser pocket it will certainly produce a fireball, possibly with dire consequences.

First turn on the oxygen valve at the cylinder, followed by the acetylene, leaving the key on the acetylene valve so that the fuel gas can be shut off quickly in case of an emergency. Turn on the acetylene valve at the torch and light the issuing gas, adjusting the flame until it is bushy. Turn it down if there is too much and it burns with a gap at the nozzle. Turn it up if there is too little and it burns with lazy rolls of soot coming off the flame. As you then slowly turn on the oxygen, the bright yellow flame will start to turn blue. Adjust the amount of oxygen until the acetylene haze clears to a well-defined lighter blue cone in the centre of the flame. This is the neutral flame that is used for welding. If there is too much oxygen, you will have

The correct spark lighter should be used to light the gas. Never use a butane gas lighter.

Turn on the acetylene at the torch and light the issuing gas.

If the gas pressure is too low, lazy rolls of soot will issue from the end of the flame. Do not leave it in this state for long as the workshop will soon be covered with falling soot particles. Turn up the pressure.

If the pressure is too high the flame will burn away from the nozzle, leaving a gap between the flame and nozzle. Turn the pressure down at the torch until the flame makes contact with the nozzle.

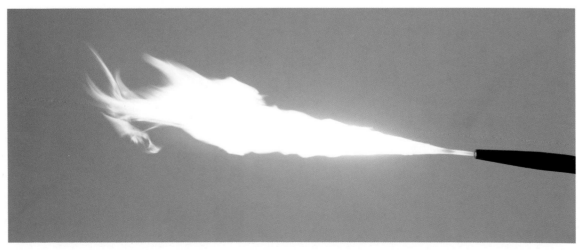

The correct pressure for the nozzle is indicated when you have a bushy flame.

As the oxygen is turned on the flame will go from orange through brilliant white to a blue haze, with a small brighter cone adjacent to the nozzle.

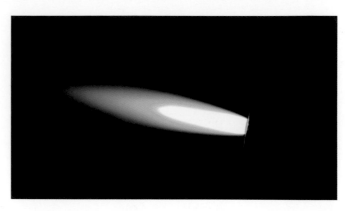

If the flame burns with an excess of acetylene the bright inner cone will have a feathery appearance.

The neutral flame is achieved when the inner cone is defined and rounded at the tip.

The oxidizing flame burns with a much sharper inner cone. It is not so bright and can hiss and pop with the excess oxygen.

an oxidizing flame, which is characterized by a much paler and more pointed inner cone, accompanied by a popping sound. The other extreme, a carburizing flame resulting from an excess of acetylene, has a longer centre cone that will not produce enough heat for welding. The uses of oxidizing and carburizing flames will be covered briefly towards the end of this chapter.

Shutting Down Procedure

The acetylene valve on the torch should be shut off first, followed by the oxygen. After the cylinder valves have been closed, release the pressure in the hoses by

opening the valves on the torch – first the oxygen and then the acetylene – until the gauges read zero. If the equipment is to be used again in the next few hours then that's it. If you are finished for the day, however, the regulator adjusters should be fully unwound.

Exception

The only exception to the above procedure is if you have a backfire into the mixing chamber of the torch. This will be evident from a roaring noise and the torch will become hot very quickly. In this case, turning the oxygen off first will extinguish the burning within the torch. Follow this quickly by turning off the acetylene. If the torch is not shut down quickly enough, the

Gas Welding Nozzle Chart

Nozzle Size	Plate Thickness mm	Plate Thickness SWG	Acetylene and Oxygen Pressure PSI	BAR	Gas Consumption Ft³/ hr	L/hr
1	0.90	20	2	0.14	1	28
2	1.20	18	2	0.14	2	57
3	2.00	14	2	0.14	3	86
5	2.60	12	2	0.14	5	140
7	3.20	10	2	0.14	7	200
10	4.00	8	3	0.21	10	280
13	5.00	6	4	0.28	13	370
18	6.25	3	4	0.28	18	520
25	8.20	0	6	0.42	25	710

Recommended pressures and gas consumption are for guidance only. Set pressure to suit job, change nozzle for next size down if pressure is too low for nozzle to prevent backfire.

Model 'O' Torch Nozzle Chart

Nozzle Size	Lead Weight lbs/ft²	Gas Consumption Ft³/hr	L/hr
1	2–3	0.175	5
2	4–5	0.425	12
3	6–8	1.100	31
4	10	2.200	62
5	18	4.500	125

mixing chamber can melt. The torch should now be left to cool off and carefully checked over before reuse.

WELDING

Now that the equipment is set up and ready to go, it is time to attempt a weld. Although it is possible to weld thick sections with gas, anything above about 3mm in thickness is uneconomic in terms of time and gas consumed. Select the right size nozzle from the table for the thickness of material being welded and set the gas pressures as described. The pieces to be joined will require clamping in position. This is even more relevant when welding with gas since the metal heats to fusion temperature more slowly, allowing more time for distortion to take place. As before, tack weld at both ends and at regular intervals along the line of the weld. This is achieved by heating the two pieces until white hot and then inserting the tip of the filler wire between them, so that the end of the filler wire melts and fuses the two pieces together. Withdraw the torch and let it cool. You should note that if the filler wire is melted just in the flame, there is a danger that the blob of molten metal will cool sufficiently before it lands on the item being welded and will not fuse properly, resulting in what is known as a cold shut.

Once tacked, the usual procedure is to start at the right-hand side and weld towards the left. For material up to 5mm, heat the tack weld until molten, using a

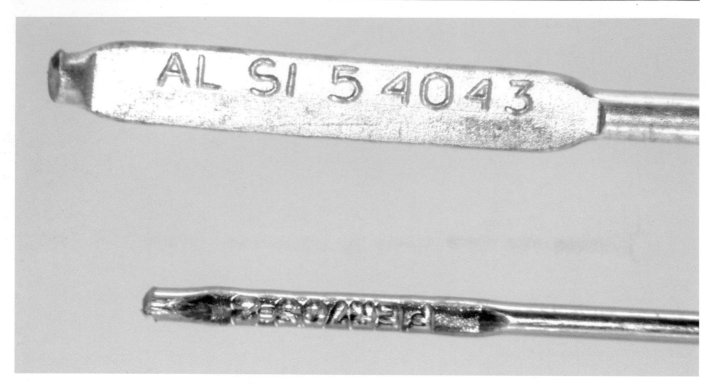

Filler rods for gas welding are stamped with the rod's composition.

slight swirling motion with the torch and holding it at about 60 to 70 degrees to the item being welded. This will ensure that both edges will be heated evenly. As a small gap, the weld pool, appears, feed in the filler rod at an angle of about 30 to 40 degrees from the opposite side to the welding torch. Maintain the swirling of the torch with the right hand and the dabbing of the filler wire with the left hand, while at the same time moving towards the left. If the weld pool appears to become too large, either increase the rate of application of the filler wire or raise the torch slightly to decrease the heat input. Do not try to weld from one side to the other all in one go. Stop after about 1in (25mm) of weld and then commence at the other end, followed by another short weld in the middle, so allowing the panel to cool down in-between, until the line of weld is complete. If the metal is melting too fast, try turning the gas down a little before carrying on, ensuring that you maintain a neutral flame. Alternatively you could go down a nozzle size.

I will repeat my earlier warnings: do not weld on

an inflammable surface and try not to use one that will conduct the heat away too quickly, such as a steel worktable. The best surface for practice welding and small gas welding jobs is one built from several fire bricks made of a heat-proof material. You must be careful, however, not to allow them to become wet or to use ordinary house bricks, since on heating the dampness contained within the brick will readily turn to steam, causing the surface of the brick to fly off at high speed. Any resulting injuries would be extremely painful on the skin, but the force would easily be enough to take out an eye. Also remember that, while the hottest part of the oxyacetylene flame is close to the nozzle tip, the flame can be in excess of 20in (500mm) long and any paper or similar in the vicinity will readily catch fire.

Practice will make you more proficient. There are many variables to be contended with in gas welding, such as nozzle sizes, gas pressures and thicknesses of filler rods, and it is this variety that makes gas welding so versatile and adaptable. Of course, a high level of hand-eye coordination is required. At first having to

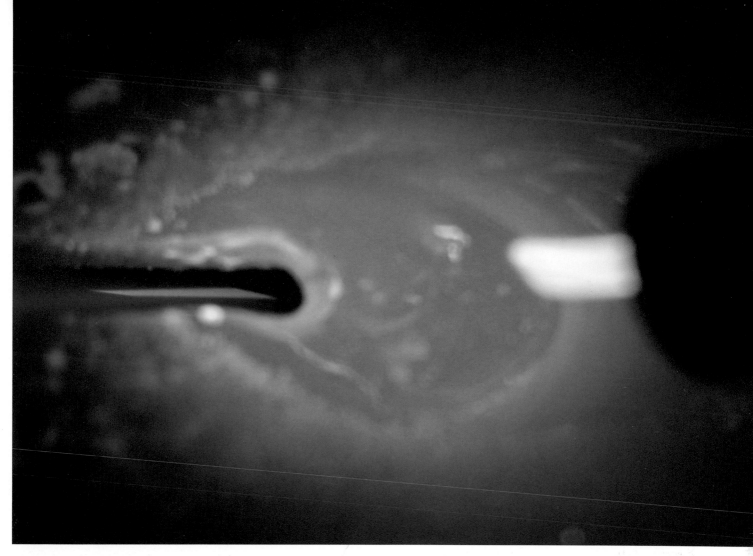

The ideal weld pool creates the shape of an onion as the surface tension of the molten metal pulls around the two parts being joined.

decide whether to turn the gas up a bit or change to the next nozzle up may seem rather daunting. After a while, though, it becomes second nature and you will soon develop a feel for what is needed.

CUTTING WITH OXYACETYLENE

Cutting ferrous metals with gas equipment is relatively straightforward once all the safety concerns have been addressed and mastered. It is only necessary to change the welding mixer on the welding torch, replacing it with a compatible cutting head that consists of an extended neck with a cutting nozzle holder on the end. These specialist nozzles have an annular ring of holes, fed with gas mixed further up the nozzle that pre-heats the metal to be cut. At the centre there is a separate single nozzle hole through which a stream of oxygen passes at high pressure when the cutting lever is opened.

The cutting process only requires the pre-heat flame to bring the steel up to about 900°C (a dull cherry red glow). It is the stream of almost pure oxygen that does the cutting, since at this temperature the oxygen combines rapidly with the steel to produce iron oxide (Fe_3O_4). Heat from the chemical reaction and the high pressure created blows the iron oxide from the kerf cleanly. At this point the pre-heat flame is no longer required to sustain the reaction, which is self-sustaining and produces enough heat on its own, but the flame continues to burn off any mill scale or other impurities ahead of the oxygen stream.

The correct angle of welding torch and filler rod are important when gas welding.

The welds achieved with gas welding can be very neat, especially if made without using filler wire.

Always remember the oxyacetylene flame is not just the hot part by the nozzle tip, as the outer flame can be 500mm long.

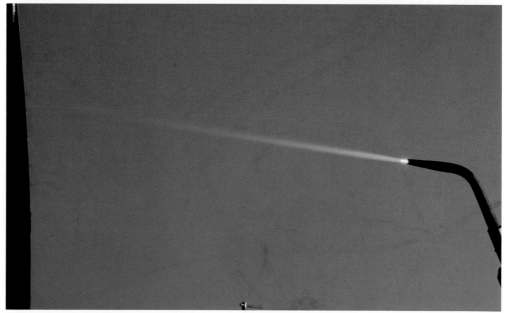

Thin sheet can obviously be cut with the standard small cutting nozzle. If cutting is something you do regularly, specialist sheet cutting nozzles are available that will cut sheet material up to 3mm, or slightly more if pushed. The advantage is that these nozzles have only two holes, one for the pre-heat flame and a second behind for the cutting oxygen stream. A slight disadvantage when using this type of nozzle is that cutting can really only be performed properly in one direction, whereas the standard cutting nozzle is omnidirectional. If used regularly, however, it will save a significant amount of gas.

Gouging nozzles are another useful addition. They can be used to gouge out untidy welds, leaving a clean kerf ready for rewelding. When thick plate has been bevelled and a multi-pass weld has been completed, the mess from a root run can be gouged out ready for a capping run on the underside. An adaptation of the gouging nozzle, the rivet cutter, can be used to cut off

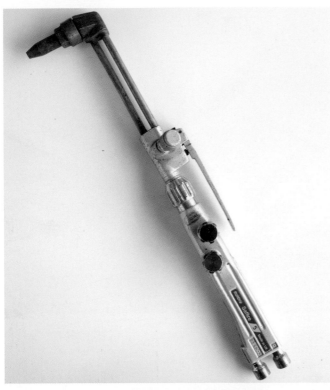

TOP: ANM cutting nozzles are used with acetylene. The three shown here will cover a cutting depth range from 3mm to 300mm.

LEFT: The standard gas welding torch can be adapted for cutting with a cutting head.

BELOW: Once the flame is adjusted to neutral, press the cutting oxygen lever and readjust back to neutral.

Preheat the plate to be cut to red heat before pressing the oxygen lever.

Once the edge of the plate is at red heat, press the oxygen lever and draw the cutting torch along steadily. Watch out for the shower of red-hot iron oxide below the cut.

If the settings are right, a neat edge will be left on the kerf.

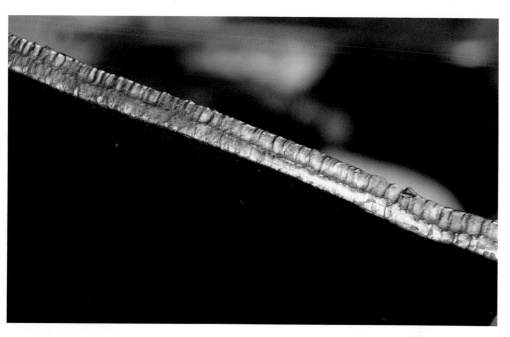

A nozzle is available for cutting thin sheet material. This will save gas, but is only suitable for use with acetylene.

Cutting can be just as satisfactorily achieved using propane as the fuel gas. Burnt with oxygen, it will reach a temperature of 2,200°C, not enough for welding but more than adequate for a pre-heat flame. More oxygen is consumed, however, as propane requires more to burn properly in order to achieve full combustion. The standard acetylene cutting nozzles (ANM) are of no use, as propane requires specially designed nozzle internals to burn correctly. Nozzles specifically for cutting with propane (PNM) that fit into the standard cutting torch or torch adapter are available in equivalent sizes to the acetylene cutting nozzles, except for the sheet cutting type.

rivets flush with the surrounding plate without cutting in, as can happen with an ordinary cutting nozzle. This is of particular benefit in boiler work in order to maintain the integral strength of the boiler.

OXYPROPANE CUTTING

As the pre-heat flame is not required to melt the metal, it follows that the high temperature from the oxyacetylene flame is not necessarily required.

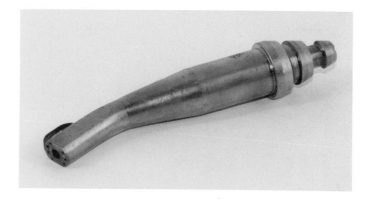

The gouging nozzle has several uses, ranging from cutting grooves in plate to cleaning back the underside of a root weld in thick plate before the final weld.

PNM nozzles, suitable for cutting with oxypropane, are able to cut the same thicknesses as the ANM nozzles, but they require more oxygen.

The PNM nozzle is made from two separate pieces as the burning characteristics of propane differ from those of acetylene.

A benefit of using propane for cutting, and indeed for heating, is that it is much more stable than acetylene and so safer to handle and transport. It is compressible, turning to a liquid at about 125lb/in^2 at normal temperatures, and is in fact sold by weight rather than volume. This means that a lot of gas can be compressed into a cylinder and is relatively cheap. As a by-product in the refining of crude oil for petrol and diesel, propane is also known as LPG (liquefied petroleum gas).

A feature of gas cutting that should be noted is that a stream of red-hot particles is produced from the underside of the item being cut and this travels a fair distance. The proper way of tackling the problem is to cut over a water tank, so that the heat from the iron oxide particles is quenched immediately. Where such an expense cannot be justified for what may be only an occasional cutting operation, take precautions and check the vicinity where you are cutting for any inflammable materials. The other danger with cutting operations is that copious amounts of carbon monoxide are released as a by-product of the formation of iron oxide. This is detrimental to health and is known as the 'silent killer'. Carbon monoxide has no smell or taste and accumulates in the body over short periods, combining with the haemoglobin in the blood and effectively starving the body's oxygen-carrying capacity.

The flame produced when cutting with propane is much the same as with acetylene, the main difference being that the sound of the flame is much harsher.

Acetylene and Propane Nozzle Mix Cutting Nozzles ANM & PNM types

Nozzle Markings	Nozzle Size	Acetylene Nozzle Mix Cuttting Nozzles ANM Gas Pressures (bar)		Propane Nozzle Mix Cutting Nozzles PNM Gas Pressures (bar)		Approx Cutting Speed (mm/min)
		Oxygen	Acetylene	Oxygen	Propane	
3–6	0.8	1.5	0.15	1.5	0.10	500–850
6–12	1.2	2.0	0.15	2.5	0.15	440–700
12–75	1.6	2.5–3.5	0.15–0.3	3.0–3.5	0.2–0.35	300–600
75–100	2.0	3.0	0.3	3.5	0.4	180–250
100–150	2.4	3.0	0.3	4.0	0.4	150–180
150–300	3.2	4.5	0.35	5.6	0.50–0.60	100–125

Gas pressures and cutting speed for guidance only.

NOZZLE MAINTENANCE

After a while when welding or cutting with gas equipment, the inevitable spitting and spluttering as detritus is encountered indicates that the nozzle needs cleaning. The usual signs are that the flame is burning to one side from the centre line of the nozzle or it is making a noise while burning. If left in this state for long there is a very real possibility of a backfire. Do not be tempted to poke about in the nozzle with any old bit of wire or a fine drill, as these will inevitably lead to a misshapen or oversized nozzle hole. The correct way to clean the nozzle is to use a reamer, or cleaner, especially designed for the job. These come as a set in a protective metal pouch that includes most of the nozzles that are likely to be used, including cutting nozzles. On opening the pouch there will be a line of wire reamers, of differing sizes, pivoted on a wire rail, and a file to reshape the end of the nozzle before selecting the right size reamer for the nozzle. It is important to keep the end of the nozzle flat and, when reaming, to keep the reamer square with the nozzle. Do not be tempted to put any side pressure on the reamer or the nozzle will become misshapen or bell-mouthed, neither of which will be good for the nozzle's performance. It will be noticed that the reamers have a rounded tip in order to avoid any corners digging in and causing further damage when inserting into the nozzle. Choosing the correct reamer is a case of selecting the one that fits into the nozzle without undue force: the size below will rattle about in the nozzle, and the one above should be too big to even contemplate using it. Once the nozzle hole is misshapen or damaged the only real solution is to purchase a new nozzle. Sometimes a large piece blocking the nozzle will be pushed further into the nozzle, and on relighting the torch, the offending piece of muck will again block the nozzle. Take the nozzle off the torch while reaming to allow the dirt to drop out from the other end. A blast from an airline should also help remove the obstruction.

BRAZING AND BRONZE WELDING

Brazing and bronze welding are in principle the same, describing different applications of the same technique. In both techniques the parent metal does not reach fusion temperature and the process is carried out at red heat, somewhere in the region of 850° to 950°C, which is well below the fusion temperature of steel. This has advantages if the full strength of a fusion weld is not required, although a brazed or bronze welded joint is strong. An oxidizing flame is advantageous. Brazing describes the joining of two or more components, using the

A spectacular shower of hot metal and sparks is produced from the underside of the plate when cutting.

Beware! Hot particles can travel quite a distance, so check for flammable materials before proceeding with any cutting operations.

Make sure you have the correct reamers for cleaning the nozzles, the right spanners for the fittings and the essential cylinder key.

Select the right size of reamer for the nozzle in hand and use the file provided to true the end of the nozzle so that it is square.

force of capillary action to draw the filler material into the joint, much the same as soldering but at a much higher temperature. Bronze welding describes the technique of using the same filler rod as brazing but building a fillet of filler material between two or more parts. The two techniques are not exclusive to one another, as is shown by the example of bronze welding a nut to a thin sheet. Where the nut is in close contact with the underlying sheet, the filler will be drawn in by capillary action. In order to further strengthen the attachment, a bronze welded fillet is built up around the nut. The strength of brazing comes from the fact that the filler material makes a bond

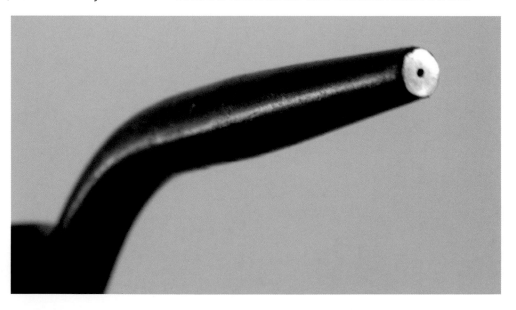

A clean nozzle.

Once cleaned, the flame should burn evenly from the nozzle. If it is at an angle or spitting and popping, the nozzle requires cleaning.

at a molecular level with the object being brazed. At extremely high magnification the surface of any material, however smooth it may look to the eye, is rough. This surface consists of the outer layer of the bonded molecules. To visualize this, think of honeycomb in three dimensions: although all the cells are packed tightly, there is a bumpy surface where they finish and each cell ends. This is what the filler material is attaching to, filling all the gaps between the molecules.

Making a Brazed Joint

As we are working at a molecular level, cleanliness is a must to get a satisfactory brazed joint. To get the capillary action to draw in the filler material, the parts to be joined must be clamped tightly together. A flux is needed to keep the joint clean while heating and carrying out the join. This has a scouring action as it melts on the surface of the parts being joined and prevents oxides forming

Flux for brazing should be mixed with a little water to form a paste before applying to the joint area and the end of the rod.

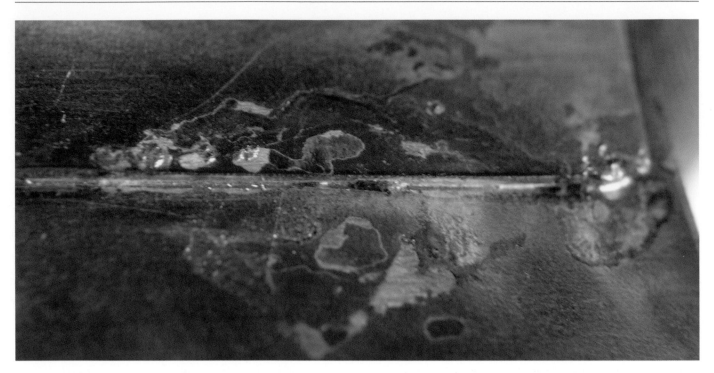

The brazed joint relies on capillary action to draw the filler rod into the joint.

that will stop the filler from adhering. Brazing rods are available that are either plain, for use with a separate flux, or have a flux coating. Sometimes flux is imbedded into dents in the rod. It is a good idea, however, to have a pot of flux handy as sometimes additional flux will be needed if the filler rod does not 'take' as the joint is made. Brazing flux comes as a powder. A general-purpose flux will suffice for general brazing duties, but more specialized types are available that cover specific temperature ranges and applications. It can be used 'as is' by dipping the end of a heated brazing rod

A bronze welded joint is produced using the same materials as a brazed joint, but with a fillet of filler material.

The correct rods and a specialist flux are important when gas welding aluminium.

into the powder, where it will adhere as a clump to the end of the rod. This clump is touched on the joint area as it is heated and it will then wick into the joint when it reaches the red heat. Reapply the rod using the heat from the joint rather than from the flame to melt it. If everything is correct the filler will be drawn into the joint. Moving further along with the heat and more filler rod will complete the joint, with a thin golden yellow line being visible. As an alternative the flux powder can be mixed with a little water to form a paste, although you must not use the original container as it will set into a hard block. Make up just enough for the job in hand and this can then be administered to the joint and put onto the rod before heating.

Making a Bronze Welded Joint

Cleanliness is just as important as it is when making a brazed joint. The same filler rods and flux are used, but a fillet is built up along the joint line instead of relying on capillary action. Once the joint is up to red heat, touching the joint surface with the fluxed rod will deposit some flux, which will be seen to spread with the heat and the surface should shine. At this stage the rod should be touched onto the surface and, if at the correct heat, this should 'tin' the clean area. An effective fillet should be produced by adding more filler rod and heat

working along the joint, adding more flux with the rod as necessary.

CLEANING UP

As a flux has been used, cleaning afterwards will be necessary. On cooling the flux will leave an extremely hard and brittle coating that will need chipping off. Goggles are essential as this slag is very much like glass and flies off at all angles in very sharp shards; it will even ping off on its own as it cools down. The hardest part to remove will be along the edge of the joint between the filler material and the parent metal, but a good scrub with a wire brush in the angle grinder should remove it.

ALUMINIUM WELDING

Aluminium welding is perfectly possible with oxyacetylene but has now been superseded by MIG and TIG welding. Even so, we will have a look at the technique involved to show how adaptable an oxyacetylene set can be.

Among the many misnomers in the field of welding, aluminium welding by gas is in fact aluminium brazing, as the filler rods melt at a lower temperature than the parent metal of the objects that we are attempting to join. Unlike steel and non-

Although it is technically brazing, welding aluminium is somewhat harder than working with steel.

ferrous metals with a higher melting temperature, aluminium melts before there is any change in the surface colour. The first sign that the metal is about to melt is a slump under the surface skin. A molten pool of metal on the floor quickly follows this, as the metal drops away.

Aluminium and its alloys are renowned for their resistance to corrosion, owing to a hard oxide layer that forms on its surface, resisting further corrosion. This oxide layer needs to be penetrated if you are to have any success in welding or brazing. In gas welding a special flux is mixed with water and applied to the parts to be joined and to the rod. Further flux can be added by dipping the rod in the flux mixture as you proceed. Once the work is finished it is imperative to wash off any residual

flux with hot water and scrub thoroughly with a brush, as the flux is acidic and will eat away at the aluminium. This is the main disadvantage of using gas to join aluminium, as any work where the reverse side is out of reach and the flux cannot be removed will rapidly corrode.

Welding Process

Any cleaning should be done with a stainless steel wire brush, as ordinary steel will contaminate the aluminium surface. As the process to be performed is actually brazing, the aluminium should not be melted. The surface, once fluxed and heated, will start to glisten as the oxide film is broken down by

All that is required for lead burning – apart from skill.

the flux. When gently applied the filler rod should now melt into the surface. If joining two pieces of sheet together, keep the flame moving evenly in a swirling motion as with steel. Proceed swiftly once the rod has started to melt to ensure that it does not melt through.

Unless you were given the specification when you bought the material, it is almost impossible to determine the alloy composition of aluminium. Various filler rods are available as pure aluminium or with varying amounts of silicon or magnesium. If a filler rod bearing magnesium is used on aluminium that does not contain that metal there is a very real possibility of cracks forming in the subsequent joint. The safest option when there is any doubt as to the make-up of the aluminium is to use a 5 per cent silicon filler rod, as this is

the most forgiving and will join most alloys, with joints relatively easily made. As an alternative, if you have any spare material to hand, a thin strip can be cut to use as a filler rod. This will ensure the filler used matches the base material and avoids any compatibility issues.

LEAD BURNING

Although still used in construction, lead is not as common in the building industry as at it once was. There is still a niche market in restoration work, however, and it is used to good effect where there is still no practical alternative to using lead.

Lead burning is included here to exemplify the multitude of uses to which gas welding equipment can

Once the weld pool forms, keep moving or the lead will be all over the floor.

be put. In case you have any fears about the poisonous properties of lead, Health and Safety guidelines state that there is little likelihood of lead poisoning being associated with lead burning provided hands are thoroughly washed after handling the metal, and that fumes do not present a danger so long as the work is carried out in an airy environment.

Lead work has traditionally been part of the plumbing trade and a number of strange terms are still used, such as 4lb lead (4 pound lead), which is the weight of the lead per square foot and equates to a sheet approximately 1.75mm thick. Although metric units have been used for many years, lead sheet is still purchased in imperial measurements.

Although the dull grey and weathered appearance of lead seems pretty mundane, it has some remarkable properties. In the periodic table lead is next to gold, and for millennia alchemists have been trying to find ways to turn lead into gold. Lead is one

of the few metals that will self-anneal at normal air temperatures, it will harden while being worked and it is resistant to most acids and other harsh chemicals in industrial processes, making it useful in the linings of chemical tanks.

The traditional title of 'lead burning' is in essence lead welding. Lead melts at a relatively low temperature of 327.5°C but conducts heat away quickly, so a hot, small flame is necessary in order to avoid a molten pool of lead on the bench. After completing the preparations listed below, welding can commence using a model 'O' welding torch with a number 2 nozzle set and a neutral flame.

To prepare the edges to be joined, all the oxide needs to be removed. This is achieved by scraping with a sharp decorator's hooked scraper until bright, shiny metal is exposed. Lead filler strips, about 3–4mm wide, are prepared by cutting them from a spare piece of lead sheet and then scraped. Examples shown here were produced on the bench using a piece of plywood to keep the two parts flat for welding together. Under those conditions it is relatively easy, but it can be a whole new ball game up on a roof and this is where the years of training come in. The two edges are placed together and tacked. Working with lead is all about confidence: any hesitation will result in a hole in the lead sheet. Once the weld pool has formed, immediately feed in the filler and work along the weld line, quickly dipping the filler in as you go. If things appear to be running away, then lift the torch and let it cool down before continuing. To master anything other than lead burning on a horizontal bench will take skill and patience. It has been included here to illustrate the flexibility of oxyacetylene equipment, and it follows that silver- and goldsmithing are within reach with a small torch, if you are so inclined.

OTHER USES

To illustrate the versatility of gas equipment, here are some more uses that are within reach, even if some are specialised. Specialist nozzles for flame cleaning produce a flat wide flame that is ideal for tasks such as paint stripping. In the classic car world, where lead loading on car bodies is still used, rather than the modern fillers, a specialist nozzle is employed that mixes the acetylene with air, rather than oxygen. This gives a lower temperature flame, which is wider and softer than the usual pointed welding flame and gives more control over the melting of the lead. Metal spraying is another option. Various metal coatings can be applied with the correct torch fitted and the right metal powders. The oxyacetylene flame heats the item to be coated and then the metal powder is introduced into the flame, where it melts in the high temperature and sticks to the item. As an example, this technique can be used to build up a worn shaft by rotating it on a lathe, at slow speed, and spraying on the metal powder to give an even layer. The advantage of this process is that a relatively soft shaft can be built up with a layer of a much harder-wearing metal compound and then ground back to the original size, giving much better wearing characteristics than the original. This is probably a simplistic overview, but gives an idea of what can be achieved with the right equipment.

HARD FACING

A carburizing flame, where the oxyacetylene flame burns with an excess of acetylene, is signified by an extended inner cone. The main use for this type of flame is for hard facing. Plough points and other items

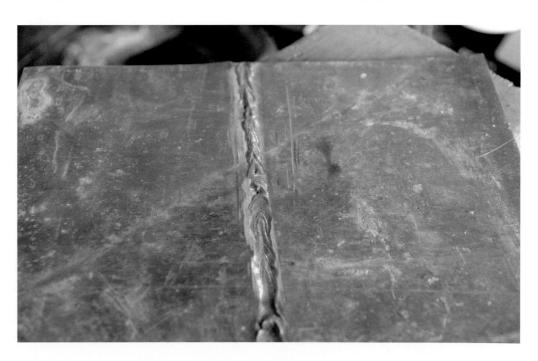

Making this sample on the flat was relatively easy, but it takes years of practice to do it up on a roof.

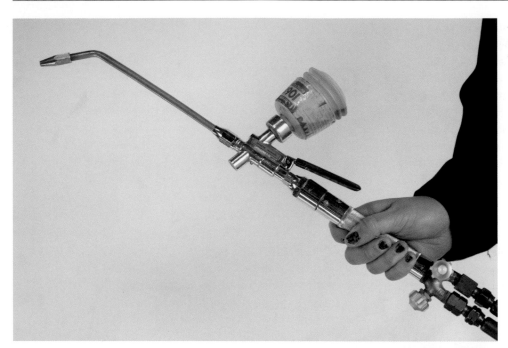

Specialist powders are available for use with a metal spraying torch to build up worn components.

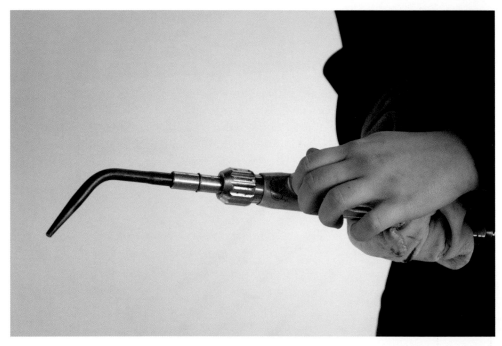

Once the job is finished, the gas should be shut off at the cylinders, the hoses purged of gas and the regulator adjusters unwound to release the spring pressure internally.

that wear rapidly can be built up and then hard faced with 'stellite' rods. These rods leave a high carbon steel layer at the surface of the item being hard faced. The purpose of the technique is to heat the surface to melting point, and not the main body of the item, so that the carbon content is not diluted through the low carbon steel. The carburizing flame, with its excess of carbon, helps keep the carbon at the surface. Once the item is up to red heat the surface will be seen to take on a wet appearance, at which point the rod should be added so that it melts into the surface.

As yet the industrial gas safety protocol of the British Compressed Gases Association's code of practice 7 (CP7) has not been fully implemented. In essence, though, it states that all gas welding equipment, including hoses, regulators and torches, must be fit for purpose. The upshot of this is that, at some time in the future, when purchasing welding gas it will be necessary to produce a certificate stating that your equipment has met the required standard. The major gas suppliers to industry run regular courses around the country on gas safety and certification. Once completed, these give you the power to certify your own gas welding equipment as well as that belonging to other people or businesses, which will help offset your costs.

It is a good idea to get into the habit of checking your equipment regularly, even if it is all set up on the trolley and seemed fine the last time that you used it. Hoses are particularly vulnerable where they rub on the rough concrete floor. Hot items abound while welding or cutting. Especially when cutting through thick plate, a shower of molten steel rains down on to the floor and it is inevitable that some will make contact with the hoses, even when the operator is being careful. When stopping for a tea break or lunch it may be enough to shut off the gas at the cylinders, but if the break is any longer you should get into the habit of shutting off at the cylinders, purging the remaining gas in the hoses and then unwinding the regulator controls to relieve the pressure on their internal components. This will ensure a longer and more accurate life. A further benefit is that, by having to set up afresh, it gives the operator a chance to actually look at the gauges on the regulator as the pressures are adjusted and make a mental note of the gas remaining in the cylinders.

It is worth reiterating that some sort of formal training should be undertaken before contemplating setting up and using your own gas welding equipment. Watching a professional using such equipment may look relatively easy, but the correct training will help you diagnose a problem long before it gets out of hand.

Oxyacetylene kit, here seen ready to go, is extremely portable and versatile.

GAS WELDING PITFALLS

High level of dexterity required to produce good welds.

Distortion can be a problem due to inefficient heat transfer during welding.

Competence required to handle equipment safely.

Fumes when cutting steel dangerous to health.

Regular checks on equipment required for continued safety.

Gas welding, although largely superseded in workshops today, has very many uses. The fillet weld and bronze welded joint demonstrated here show how adaptable this type of equipment is in the workshop.

Fillet Weld Using Gas Welding Equipment

1. Adjust the flame at the regulators to give a neutral flame.

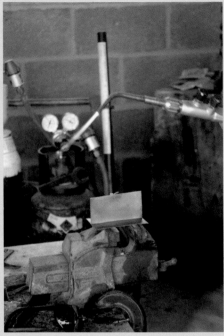

2. Clamp the parts to be welded.

3. With the filler rod in your left hand and the gas torch in the right, tack weld at the ends and at regular intervals along the joint.

4. Heat both edges, but concentrate more heat on the lower one, using a circular motion, and then introduce the filler wire.

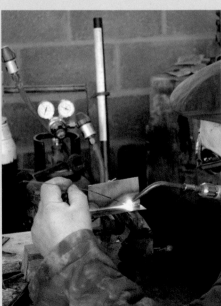

5. Weld between the tack welds, ensuring that they are melted into the weld pool to avoid any cold shuts. Using a weaving motion with the gas torch, dab the filler wire into the weld pool, but don't let it melt in the flame.

6. The finished gas welded fillet weld.

Bronze Welded Joint

1. An oxidizing flame, used for brazing and bronze welding, helps to prevent the zinc in the filler rod from boiling off.

2. Flux can be mixed to a paste and applied to the joint, or flux coated rods can be used.

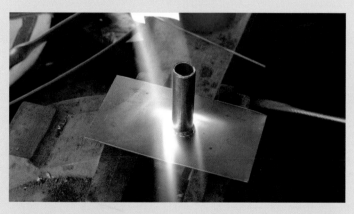

3. Heat the whole joint evenly, keeping the flame moving. Note the flux on the filler rod (right).

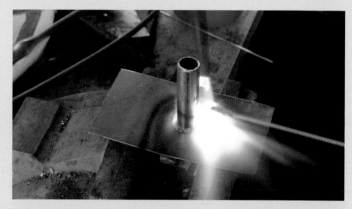

4. As the parts become red hot, the flux will be seen to spread across the surface. Apply the rod to the hot surface, not into the flame, and the filler will be seen to flow into the joint.

5. For a bronze welded joint, add more filler until a fillet builds up. Keep moving the flame around the joint so that the filler flows evenly.

6. The completed joint. Mild steel components can be quenched once they have cooled below red heat. This will help to crack off the flux, which sets as hard as glass. Chipping and wire brushing will remove any remaining flux.

8 Spot or Resistance Welding

SPOT WELDING

Easy technique to use, but has limited applications.

Main applications
- Automotive body panels, mild and stainless steel sheet work.
- Clean welds produced with little preparation before painting.
- No consumables required.
- Just need to keep electrode shape for good quality weld.

Although spot welding has a somewhat limited use in a general workshop, for completeness it will be looked at here.

The spot welder uses electricity as the operating medium, but does so in a way completely unlike that of the usual arc welding techniques. Heat for the arc welder in all its forms – manual arc, MIG or TIG – is produced from the arc between the electrode and the item being welded, and then through the earth return lead to complete the circuit. The spot welder relies on the resistance that exists between the two surfaces to be joined, with the lap joint being the only practical application of this technique. The spot welding machine in its simplest form is a transformer to bring down the voltage and increase the current available, which is directed through two copper electrodes. As these two electrodes close and squeeze the two pieces together, a built-in switch turns on the current from

The spot welder is rather specialist in its use, but it can be useful if much sheet metalwork is envisaged.

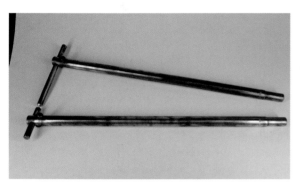

TOP: *A spot welder is easy to use – just squeeze the handles together for a few seconds.*

BOTTOM: *Different arms are available for differing jobs, but they are expensive as they are made mainly from copper.*

RIGHT: *Spot welds require no remedial cleaning once produced.*

the transformer. While the current flows through the circuit formed by the transformer and copper electrodes, the items being squeezed together present a higher resistance than the rest of the circuit. This resistance to current flow produces

The only maintenance required is to maintain the tips in the correct profile.

heat, which is enough in the small spot between the electrodes to melt the surfaces of the two components and, with the pressure being applied between the electrode arms, fuses the two parts together at that point.

This very simple process, although limited in its uses, has many advantages for construction of sheet metal components. The most obvious is in the car construction industry, where robots can spot weld a car shell together in no time. Other distinct advantages are that no flux or shielding gas is required and ultraviolet radiation is not produced, except perhaps for the occasional flash from the contact points of the copper electrodes. Being localized, the total heat input is greatly reduced and so less distortion will be generated from this process.

The use of resistance welding has many uses apart from producing a spot-weld. By changing the simple copper electrodes for rotating contact wheels, the spot can become a continuous weld for producing watertight seams suitable for tanks and so on. The seam welder is a lot more complex than the spot welder, but the underlying principles are the same.

The handheld spot welder is a far cry from the robotic leviathans seen in factories. These can weld almost continuously owing to channels running through the copper electrodes, through which cooling water flows to take away the excess heat as it dissipates from the spot weld into the copper tips. At lower levels, if anything more than two bits of flat sheet are to be joined, then the standard electrodes will be of little use, especially if the items are in more than one plane. Electrodes come in differing shapes to complete welds in a variety of positions. Since they are made from copper, or copper alloys, and of necessity are built from a large cross-section, they are correspondingly expensive and have just that limited use.

Apart from the initial outlay and the purchase of alternative arms, the running costs of a spot welder are low: only the electricity going through the machine has to be paid for and there are no filler wires or electrodes to purchase at regular intervals. Maintenance is minimal. The only real work to do is to check the ends of the copper electrodes. Keeping these to their original profile will maintain the welder's

efficiency and its ability to make a strong and secure weld.

ALTERNATIVES

As mentioned in Chapter Five, a very good replacement for a spot weld is a plug weld made with a MIG welder; the two sheets are arranged as a lap joint and the weld is made through holes punched in the top sheet. The most obvious advantage is that there is no need to buy a specialist machine. In addition, since you need access to only one side of the panel, it can be a boon to the car restorer. Special MIG welding shrouds have been developed to replace the standard ones when plug welding. These make it easier to carry out the plug weld accurately, as the shroud is pushed against the top panel, centralizing the torch over the hole in the top sheet and holding the two sheets to be joined in close proximity. To allow this to happen, the shroud has notches around the periphery to allow the shielding gas to escape as the weld is made. To make a shroud suitable for plug welding, all that is required is to cut two notches in a spare shroud. Used in conjunction with a modern

Shrouds are available specifically for plug welding, but it is easy enough to make one from a spare shroud.

Finished plug welds, if done correctly, give a neat appearance to the job.

weld-through primer, the method has put an end to the old problem associated with spot welds that the sheets corroded away where the panels were devoid of paint. A development of the standard plug weld is that, rather than just punching a round hole in the top panel, a short slot can be produced with the punch by overlapping the holes. Once this slot is welded as a plug weld the result is very strong.

machine is available that will spot weld two sheets of 2mm sheet but, despite also being 240 volts, it requires its own dedicated power supply.

If a lot of spot welding is envisaged, floor-standing machines are available, but at a price. They also require a substantial power supply and they are really only for production work in a factory environment.

WHAT'S AVAILABLE

The smallest handheld machine that will typically weld together two 1mm pieces of sheet requires a minimum of 16 amp 240 volt supply, giving a high welding current for a short duration, and will have a duty cycle of around three spot welds per minute. A better version is fitted with a built-in adjustable timer so that every weld comes out the same once the correct time per weld is found. This bypasses the guesswork of holding the welder switched on while guessing if this weld is as long as the last. A larger capacity handheld

SPOT WELDING PITFALLS

Limited uses.

Equipment relatively expensive to purchase.

Without post-welding treatment, corrosion between panels can be a problem.

Spot welding is a very useful, if limited technique. It is demonstrated here on stainless steel, but its main use is on mild steel panels. The plug weld is demonstrated as an alternative to a spot weld. Performed with a MIG welder, it has the advantage that access is needed to only one side of the panel, and it demonstrates the versatility of the MIG welder in contrast to the limitations of the spot welder.

Spot Welded Seam

1. For spot welds to be successful, all surfaces need to be clear of contamination.

2. Clamps should be positioned to hold the two parts tightly together.

3. Spot weld between the clamps.

4. Reposition the clamps if necessary to maintain close contact along the joint.

5. Remove the clamps and fill in with spot welds along the joint, evenly spaced between 5 and 10 mm apart.

6. The finished joint needs very little, if any, post-welding treatment.

Plug Welded Seam

1. To produce a plug welded joint, either drill holes in one sheet or, preferably, use a punch that leaves clean, burr-free holes.

2. Use clamps to hold the two parts together tightly.

3. A finished plug weld should be neat and tidy, leaving very little distortion.

4. The underside of the plug weld, showing complete penetration for a strong weld.

5. Using the correct nozzle on the MIG torch allows easy alignment of the torch over the punched hole.

6. The finished job: welds can be left as they are or ground back, depending on whether it matters that they show.

9 Choosing the Right Equipment

Talking about the theory and practicalities of the welding processes is all very well, but where and what to purchase is a completely different matter.

The old adage 'you get what you pay for' is certainly correct with welding equipment. A cheaper model with few features will possibly suffice if your welding requirements are fairly modest, but if you want to weld all day and every day a more expensive model may work out cheaper in the long run. Later in this chapter I will explain about the all-important duty cycle in more detail, but briefly, the better the duty cycle the longer you can weld without having to allow the machine to cool down.

As an example, if your main interest is mmA welding but you need to TIG weld together some mild steel items, it would make sense to purchase an inverter welder that covers the stick welding requirements. Purchasing a TIG welding kit to attach to the inverter welding machine gives you a basic scratch start DC TIG welding set-up. If you anticipate working with aluminium that requires TIG welding, however, you might as well purchase a dedicated TIG welder from the start. Most of the dedicated TIG welding machines are capable of stick welding, but they also come with all the luxuries offered by TIG welding, such as various trigger settings. Instead of the hold and weld type found on MIG welders, which can be tiring as you have to hold the trigger all the time you are welding, with this refinement the trigger is pulled once to switch on the current and again to switch off. Perhaps this may sound a bit extravagant for a few welds, but such enhancements will pay dividends if you are welding all day long.

If your main requirements for welding are car bodywork and thin sheet steel, you will find that

Some of the kit available today: (left to right) MMA inverter welder; small MIG welder with throwaway gas bottle; larger MIG welder with large gas cylinder; AC arc welder; and a spot welder in the centre foreground.

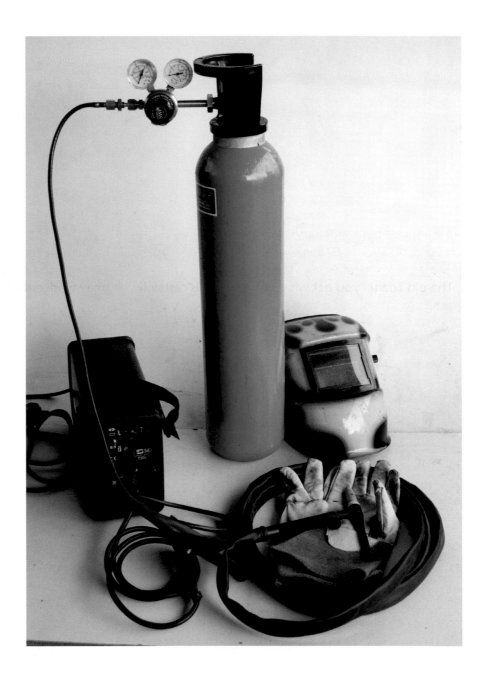

An MMA inverter welder set up for TIG welding with an add-on kit.

although the TIG welding apparatus can cope admirably with thin section steel, it is slow when compared with a MIG welding machine, which welds together car body panels at a reasonably fast pace when set up correctly, with little fuss and distortion. The MIG welder can also be used for general fabrication, and aluminium and stainless steel can be tackled with a change of filler wire and the correct shielding gas. If you can get a reel of brazing wire, the MIG will braze a multitude of jobs, such as joining zinc-plated steel sheet. Modern

car body panels also require brazing, so as not to destroy the properties of the boron-bearing steel panels.

COSTS

If you ignore the initial costs of purchasing welding equipment, the MIG welder wins hands down when it comes to running costs per unit length of weld, even when the costs of the shielding gas are taken into

An MIG welder set-up with a 'Y' size shielding gas cylinder is ideal for trundling around the workshop on a purpose-built trolley.

consideration. As stated in Chapter Five, the initial costs of buying larger quantities of filler wire and shielding gas are higher, but the costs pro rata are very much reduced.

Arc welding rods are readily available for a multitude of tasks, including such specialist ones as hard facing and welding cast iron, but they can be relatively expensive, with the small 1kg packs costing well into double figures. This has to be borne in mind when considering the expense of replacing a broken casting with a new one: in comparison welding rods might not seem so expensive, making repair a cost-effective option.

WHERE TO BUY EQUIPMENT

There are many choices when it comes to buying equipment. Several national chains of tool shops can supply by mail order and some, but not all, give detailed descriptions and comparisons of the models that they carry. Information can be found on the internet about almost every make or model, and dedicated online forums discuss the merits and problems associated with specific models. Purchases can be made directly online and better discounts may be had that way. The internet is a fine reference source but it is always a good idea to actually see the machine that you require.

Specialist welding suppliers are dotted around the country and will be able to recommend a particular model for your requirements. Their staff are usually welders themselves and have possibly come from the welding industry, so they will be very knowledgeable in their particular field. It would not be in their interest to sell an unsuitable machine, as they trade on their reputations. The specialist will also be able to give backup with spares and further advice, if it is required. The final price for this enhanced service may well be higher than that charged by a larger tool company. In the event of a problem with your welding machine, however, the usual response from the large tool companies is to try and sell you a new machine, as they have little knowledge of the products they sell.

SHIELDING GAS SUPPLIES

It was made clear in Chapter Five that the larger cylinders of shielding gas contain more gas pro rata. The small throwaway cylinder is an extremely expensive way to buy your shielding gas, although it may be cost-effective to buy a small cylinder of pure argon if you have only a small job to do, perhaps in aluminium, and are unlikely to do any more in the future.

Until recently the only other option was to open an account with one of the national companies that bottle the gas. If a lot of welding is envisaged the annual, three- or five-year rental schemes for their smaller-sized cylinders begin to make sense. Once you have an account and require a different type of gas for a short period, a small cylinder can be rented on a monthly basis. The opposite holds true if you require a large amount of gas for a big job, as the full-sized industrial cylinders are available on the monthly tariff.

Several vendors with branches around the country have recently introduced ranges of hobby gas cylinders. Instead of renting the bottle for a recurring fee, as in an account-based system, you pay a one-off deposit (£55.00 at the time of writing) plus the cost of the gas and delivery. From then on, unless you require a second bottle, you pay for just the gas and delivery, however long you keep the cylinder. Once you have decided to give up the cylinder, the deposit will be reimbursed to the original hirer on return of the cylinder. This is rather a win–win situation for the small to medium user. However long you take to use the gas in the cylinder, whether it is a month or several years, there are no more costs involved. The hobby gas type of cylinders varies from the industrial

Shielding gas cylinders are available to rent from various sources.

When carrying gas cylinders in a vehicle, comply with regulations by displaying the correct safety signs.

standard in having the gas outlet at the side of the valve block, rather than out of the top. The thread is standard and the standard regulator will fit, but unless the regulator is positioned on its side, or at an angle from the vertical, the adjustment knob on the regulator hits the cylinder handle/valve safety guard. This makes it difficult to read the content gauges on the regulator, and the adjustment mechanism within the regulator may possibly be damaged if the cylinder connection nut is tightened while the two are in contact. An alternative type of regulator is manufactured for these side-exit cylinders that clears everything when fitted and the content dials are the right way up for easy reading.

While on the subject of gas regulators, the type with two gauges, showing the pressure remaining in the cylinder as well as the amount being used, is well worth the investment of a few pounds extra. The slightly cheaper ones with only one gauge, showing what is being used, are acceptable but it is very frustrating if they run out of gas unexpectedly at the weekend.

ROAD TRAFFIC ACT COMPLIANCE

When purchasing shielding gas and or oxy-fuel gas from a major industrial supplier you can opt to have your supplies delivered. This can work out rather expensive on top of the price of the gas. Collection is an alternative, but you must comply with the terms of the Road Traffic Act. This entails displaying the correct signs on your carrying vehicle, either a red diamond with a graphic of flames and the words 'Flammable Gas' for fuel gases such as acetylene and propane, or a green diamond for non-flammable gases, such as argon, carbon dioxide and oxygen, with the words 'Non Flammable Gas' or 'Compressed Gas' and the graphic of a gas cylinder. The signs are available as stickers from gas suppliers, so there is no excuse not to comply. Nobody wants to contemplate a car accident, but in the event of an accident these signs tell the emergency services that your vehicle is carrying potentially lethal cylinders, especially if fire is involved. Once transportation of the cylinders is complete, remove the stickers until required again, as otherwise

1~ ——○○——1~			EN 60974-6:2003			
Uo: 48V	~ 50 Hz		I2: 55A/20.2V – 160A/24.4V			
Ø mm			2.0	2.5	3.2	4.0
I2: A			55	80	115	160
tw (S)			470	211	106	68
tr (S)			892	805	727	685
U1: 230V	22A	I1: max 36.8A I1: eff 11.8A				
1~50 Hz	IP21	(K=1.2)	16.2 Kg			

The duty cycle label gives all the information you require for the particular machine.

DUTY CYCLE

If you intend doing any serious amounts of welding, then consideration needs to be given to the duty cycle. This will be marked on the specification plate, which should be fitted somewhere on the machine, and will appear as a percentage, either singly as a percentage of the machine's full power, or several at various power settings, for example 40 per cent @ 150 amps, 65 per cent @ 100 amps and so on.

display of the signs could potentially create a situation at an accident where emergency medical treatment is delayed while non-existent gas cylinders are being looked for.

Full or empty cylinders are heavy, so make sure that they are secured properly in the vehicle as a loose cylinder may have lethal consequences if you have to brake severely. Do not forget that all gases are harmful, so drive with a window open.

The Ten-Minute Rule

These figures may seem thoroughly confusing at first, but their meaning is fairly simple once grasped. Let's take a 50 per cent duty cycle as an example. The duty cycle of the machine is based on a ten-minute cycle of use. A 50 per cent duty cycle simply means that, in any ten minutes, the welder can be used at the specified current for 50 per cent of that period, giving a maximum of five minutes welding time. Similarly, a 65 per cent duty cycle would give 6½ minutes of welding time in ten minutes and so on.

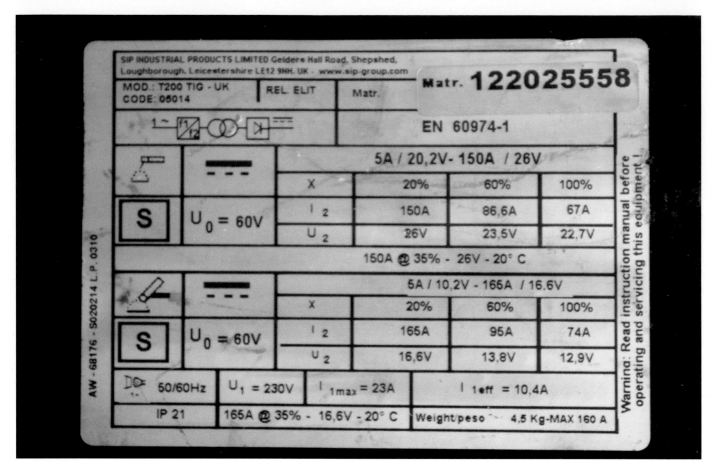

Another example of a duty cycle.

This has come about as manufacturers attempt to keep down costs and the weight of their machines. Arc and MIG welders employ a transformer to convert the mains voltage to a lower level and increase the amperage for welding. This transformer consists of a laminated iron core with wire windings of copper, or aluminium on cheaper models, that are covered with a shellac varnish to insulate the wires from each other. Any work performed results in heat being generated. In this case the resistance in the windings will create heat as the electrons flow around the coils. This unwanted heat has to go somewhere and is absorbed by the iron core before being dissipated to the air inside the machine. To increase the efficiency of heat dissipation, manufacturers fit a fan to blow across the transformer in a bid to increase the duty cycle. (The grandiose label of

Turbo is added to the machine's title.) To prevent too much heat damaging the wire insulation inside the transformer, a heat-sensitive switch is fitted that trips if the temperature of the windings rises above a safe level, cutting power to the machine until it has cooled sufficiently.

CHOOSING A MACHINE

If you do not intend to carry out more than just the occasional job lasting longer than a few minutes, then any advertised welder at the power you require will suffice. Any plans for more serious welding, however, will require caution when choosing a machine.

The model numbers that manufacturers give to their machines sometimes give the false impression that the highest amperage is more than they can actually put out. A machine called a 'MIG 225 XXX', for example, may have a headline amperage of 200 amps, but further down the specification you will find the duty cycle, typically 60 per cent. Do not forget that this really means six minutes of welding in every ten at 123 amps, which is only just more than half the power available. At the top end the duty cycle will be little more than a minute of welding, if you are lucky. At the lower amperages required for welding thin material, such a car body panels, the duty cycle may perhaps be approaching 100 per cent, which means you can weld all day long without waiting for the machine to cool down. The other figure to look for in the specifications is the minimum amperage that the machine can go down to. This comes into play if you want to weld thin sheet material, since

the arc will be more stable at a lower amperage and there is less risk of blowing clean through the panel being worked on. It is the usual story that the more you pay for your equipment, the better and easier it will be to use.

With arc welders the current output is obviously important and the duty cycle parameters apply to the welding situation. If anything more than general-purpose welding is to be done, such as hard facing, the OCV (open circuit voltage) value will then become an issue as specialist rods require a higher OCV than general-purpose rods. The requirements for the satisfactory use of arc welding rods, which are made to codified industrial standards for specific uses, are invariably printed on the original packaging label.

Useful Addresses

The following list of useful addresses where equipment and safety gear can be purchased and information sought is only a starting point. I have no connection with any companies mentioned.

Machine Mart Limited
211 Lower Parliament Street, Nottingham
NG1 1GN
Telephone: 0844 880 1250
www.machinemart.co.uk

Products can be viewed and purchased at about sixty branches around the country. The company publishes an extensive colour catalogue and goods may also be purchased online.

Northern Tool + Equipment
Anson Road, Portsmouth PO4 8SX
Telephone: 0845 605 2266
www.NorthernToolUK.com

Frost Auto Restoration Techniques Limited
Crawford Street, Rochdale, Lancashire
OL16 5NU
Telephone: 01706 758 258
www.frost.co.uk

The company supplies restoration tools mainly for the classic car enthusiast, but also stocks some useful tools for the budding welder that are hard to find.

BOC
BOC Customer Services Centre
PO Box 6, Priestley Road, Worsley, Manchester
M28 2UT
Telephone: 0800 111 333
www.boconline.co.uk

Nationwide network of branches supplying industrial gases and welding equipment. An account is required to purchase gases, but ancillary equipment is sold over the counter to non-account holders.

Air Products PLC
2 Millennium Gate, Westmere Drive, Crewe,
Cheshire CW1 6AP
Telephone: 0800 389 0202
www.airproducts.co.uk

Hobbyweld Gas
Dixons of Westerhope Limited, Westfield,
New Biggin Lane, Westerhope, Newcastle upon Tyne
NE5 1LX
Telephone: 0800 433 4331
www.hobbyweld.co.uk

The company supplies a nationwide chain of retailers that supply shielding gases and oxygen on payment of a deposit for the first cylinder, after which the only payment is for the gas purchased, no matter how long you have the cylinder. The original deposit payer can reclaim the deposit on return of the cylinder.

Index